Pre-Algebra Essentials FOR DUMMIES®

by Mark Zegarelli
with Krista Fanning

Wiley Publishing, Inc.

Pre-Algebra Essentials For Dummies®

Published by
Wiley Publishing, Inc.
111 River St.
Hoboken, NJ 07030-5774
www.wiley.com

Published by Wiley Publishing, Inc., Indianapolis, Indiana

Published simultaneously in Canada

Library of Congress Control Number: 2010924584

ISBN: 978-0-470-61838-7

Manufactured in the United States of America

10 9 8 7 6 5 4

About the Authors

Mark Zegarelli is the author of *Logic For Dummies* (Wiley) plus three *For Dummies* books on pre-algebra and Calculus II. He holds degrees in both English and math from Rutgers University. Mark lives in Long Branch, New Jersey, and San Francisco, California.

Krista Fanning writes and edits textbooks and supplementary materials for several publishing houses. As a former elementary school teacher, she has a passion for education and details. In her publishing career, she has been involved in the production of over 50 titles. She enjoys spending time with her family and stalking her local library for the newest mysteries and thrillers.

Publisher's Acknowledgments

We're proud of this book; please send us your comments through our Dummies online registration form located at `http://dummies.custhelp.com`. For other comments, please contact our Customer Care Department within the U.S. at 877-762-2974, outside the U.S. at 317-572-3993, or fax 317-572-4002.

Some of the people who helped bring this book to market include the following:

Acquisitions, Editorial, and Media Development

Senior Project Editor: Tim Gallan

Acquisitions Editor: Lindsay Lefevere

Senior Copy Editor: Danielle Voirol

Technical Reviewers: David Herzog, Amy Nicklin

Editorial Program Coordinator: Joe Niesen

Editorial Manager: Michelle Hacker

Editorial Assistants: Jennette ElNaggar, David Lutton, Rachelle Amick

Cover Photo: © iStock / Alistair Forrester Shankie

Cartoons: Rich Tennant (`www.the5thwave.com`)

Composition Services

Project Coordinator: Sheree Montgomery

Layout and Graphics: Carrie A. Cesavice, Joyce Haughey, Ronald G. Terry

Proofreaders: Melanie Hoffman, Sossity R. Smith

Publishing and Editorial for Consumer Dummies

Diane Graves Steele, Vice President and Publisher, Consumer Dummies

Kristin Ferguson-Wagstaffe, Product Development Director, Consumer Dummies

Ensley Eikenburg, Associate Publisher, Travel

Kelly Regan, Editorial Director, Travel

Publishing for Technology Dummies

Andy Cummings, Vice President and Publisher, Dummies Technology/General User

Composition Services

Debbie Stailey, Director of Composition Services

Contents at a Glance

Contents

Introduction

Why do people often enter preschool excited about learning how to count and leave high school as young adults convinced that they can't do math? The answer to this question would probably take 20 books this size, but solving the problem of math aversion can begin right here.

Remember, just for a moment, an innocent time — a time before math inspired panic attacks or, at best, induced irresistible drowsiness. In this book, I take you from an understanding of the basics to the place where you're ready to enter any algebra class and succeed.

About This Book

Somewhere along the road from counting to algebra, most people experience the Great Math Breakdown. Please consider this book your personal roadside helper, and think of me as your friendly math mechanic (only much cheaper!). The tools for fixing the problem are in this book.

I've broken down the concepts into easy-to-understand sections. And because *Pre-Algebra Essentials For Dummies* is a reference book, you don't have to read the chapters or sections in order — you can look over only what you need. So feel free to jump around. Whenever I cover a topic that requires information from earlier in the book, I refer you to that section or chapter in case you want to refresh yourself on the essentials.

Note that this book covers only need-to-know info. For a broader look at pre-algebra, you can pick up a copy of *Basic Math & Pre-Algebra For Dummies* or the corresponding workbook.

Conventions Used in This Book

To help you navigate your way through this book, I use the following conventions:

- *Italicized* text highlights new words and defined terms.
- **Boldfaced** text indicates keywords in bulleted lists and the action part of numbered steps.
- `Monofont` text highlights Web addresses.
- Variables, such as x and y, are in italics.

Foolish Assumptions

If you're planning to read this book, you're likely

- A student who wants a solid understanding of the core concepts for a class or test you're taking
- A learner who struggled with algebra and wants a reference source to ensure success in the next level
- An adult who wants to improve skills in arithmetic, fractions, decimals, percentages, geometry, algebra, and so on for when you have to use math in the real world
- Someone who wants a refresher so you can help another person understand math

My only assumption about your skill level is that you can add, subtract, multiply, and divide. So to find out whether you're ready for this book, take this simple test:

$5 + 6 = \_\_$

$10 - 7 = \_\_$

$3 \times 5 = \_\_$

$20 \div 4 = \_\_$

If you can answer these four questions, you're ready to begin.

Icons Used in This Book

Throughout the book, I use three icons to highlight what's hot and what's not:

This icon points out key ideas that you need to know. Make sure you understand before reading on! Remember this info even after you close the book.

Tips are helpful hints that show you the quick and easy way to get things done. Try them out, especially if you're taking a math course.

Warnings flag common errors that you want to avoid. Get clear about where these little traps are hiding so you don't fall in.

Where to Go from Here

You can use this book in a few ways. If you're reading this book without immediate time pressure from a test or homework assignment, you can certainly start at the beginning and keep on going through to the end. The advantage to this method is that you realize how much math you *do* know — the first few chapters go very quickly. You gain a lot of confidence as well as some practical knowledge that can help you later on, because the early chapters also set you up to understand what follows.

Or how about this: When you're ready to work, read up on the topic you're studying. Leave the book on your nightstand and, just before bed, spend a few minutes reading the easy stuff from the early chapters. You'd be surprised how a little refresher on simple stuff can suddenly cause more-advanced concepts to click.

If your time is limited — especially if you're taking a math course and you're looking for help with your homework or an upcoming test — skip directly to the topic you're studying. Wherever you open the book, you can find a clear explanation of the topic at hand, as well as a variety of hints and tricks. Read through the examples and try to do them yourself, or use them as templates to help you with assigned problems.

The 5th Wave By Rich Tennant

"David's using algebra to calculate the tip. Barbara– would you mind being a fractional exponent?"

Chapter 1

Arming Yourself with Math Basics

In This Chapter

- Identifying four important sets of numbers
- Reviewing addition, subtraction, multiplication, and division
- Examining commutative, associative, and distributive operations
- Knowing exponents, roots, and absolute values
- Understanding how factors and multiples are related

You already know more about math than you think you know. In this chapter, you review and gain perspective on basic math ideas such as sets of numbers and concepts related to the Big Four operations (adding, subtracting, multiplying, and dividing). I introduce you (or reintroduce you) to properties and operations that will assist with solving problems. Finally, I explain the relationship between factors and multiples, taking you from what you may have missed to what you need to succeed as you move onward and upward in math.

Understanding Sets of Numbers

You can use the number line to deal with four important *sets* (or groups) of numbers. Each set builds on the one before it:

- **Counting numbers (also called natural numbers):** The set of numbers beginning 1, 2, 3, 4, ... and going on infinitely
- **Integers:** The set of counting numbers, zero, and negative counting numbers

- **Rational numbers:** The set of integers and fractions
- **Real numbers:** The set of rational and irrational numbers

Even if you filled in all the rational numbers, you'd still have points left unlabeled on the number line. These points are the irrational numbers.

An *irrational number* is a number that's neither a whole number nor a fraction. In fact, an irrational number can only be approximated as a *non-repeating decimal.* In other words, no matter how many decimal places you write down, you can always write down more; furthermore, the digits in this decimal never become repetitive or fall into any pattern. (For more on repeating decimals, see Chapter 5.)

The most famous irrational number is π (you find out more about π when I discuss the geometry of circles in Chapter 11):

$$\pi = 3.14159265358979323846264338327950288419716939937510...$$

Together, the rational and irrational numbers make up the *real numbers,* which comprise every point on the number line.

The Big Four Operations

When most folks think of math, the first thing that comes to mind is four little (or not-so-little) words: addition, subtraction, multiplication, and division. I call these operations the *Big Four* all through the book.

Adding things up

Addition is the first operation you find out about, and it's almost everybody's favorite. Addition is all about bringing things together, which is a positive thing. This operation uses only one sign — the plus sign (+).

When you add two numbers together, those two numbers are called *addends,* and the result is called the *sum.*

Adding a negative number is the same as subtracting, so 7 + –3 is the same as 7 – 3.

Take it away: Subtracting

Subtraction is usually the second operation you discover, and it's not much harder than addition. As with addition, subtraction has only one sign: the minus sign (–).

When you subtract one number from another, the result is called the *difference.* This term makes sense when you think about it: When you subtract, you find the difference between a higher number and a lower one.

Subtracting a negative number is the same as adding a positive number, so 2 – (–3) is the same as 2 + 3. When you're subtracting, you can think of the two minus signs canceling each other out to create a positive.

Multiplying

Multiplication is often described as a sort of shorthand for repeated addition. For example,

4 × 3 means *add 4 to itself 3 times:* 4 + 4 + 4 = 12

9 × 6 means *add 9 to itself 6 times:* 9 + 9 + 9 + 9 + 9 + 9 = 54

When you multiply two numbers, the two numbers that you're multiplying are called *factors,* and the result is the *product.* In the preceding example, 4 and 3 are the factors and 12 is the product.

When you're first introduced to multiplication, you use the *times* sign (×). However, algebra uses the letter *x* a lot, which looks similar to the times sign, so people often choose to use other multiplication symbols for clarity.

Arriving on the dot

In math beyond arithmetic, the symbol · replaces ×. For example,

6 · 7 = 42 means 6 × 7 = 42

53 · 11 = 583 means 53 × 11 = 583

That's all there is to it: Just use the · symbol anywhere you would've used the standard times sign (×).

Speaking parenthetically

In math beyond arithmetic, using parentheses *without* another operator stands for multiplication. The parentheses can enclose the first number, the second number, or both numbers. For example,

$3(5) = 15$ means $3 \times 5 = 15$

$(8)7 = 56$ means $8 \times 7 = 56$

$(9)(10) = 90$ means $9 \times 10 = 90$

However, notice that when you place another operator between a number and a parenthesis, that operator takes over. For example,

$3 + (5) = 8$ means $3 + 5 = 8$

$(8) - 7 = 1$ means $8 - 7 = 1$

Doing division lickety-split

The last of the Big Four operations is division. *Division* literally means splitting things up. For example, suppose you're a parent on a picnic with your three children. You've brought along 12 pretzel sticks as snacks and want to split them fairly so that each child gets the same number (don't want to cause a fight, right?).

Each child gets four pretzel sticks. This problem tells you that

$12 \div 3 = 4$

As with multiplication, division also has more than one sign: the *division sign* (÷) and the *fraction slash* (/) or fraction bar (—). So some other ways to write the same information are

$^{12}/_{3} = 4$ and $\frac{12}{3} = 4$

When you divide one number by another, the first number is called the *dividend,* the second is called the *divisor,* and the result is the *quotient.* For example, in the division from the

earlier example, the dividend is 12, the divisor is 3, and the quotient is 4.

Fun and Useful Properties of the Big Four Operations

When you know how to do the Big Four operations — add, subtract, multiply, and divide — you're ready to grasp a few important properties of these important operations. *Properties* are features of the Big Four operations that always apply no matter which numbers you're working with.

Inverse operations

Each of the Big Four operations has an *inverse* — an operation that undoes it. Addition and subtraction are inverse operations because addition undoes subtraction, and vice versa. In the same way, multiplication and division are inverse operations. Here are two inverse equation examples:

$184 - 10 = 174$ $\quad$ $4 \cdot 5 = 20$

$174 + 10 = 184$ $\quad$ $20 \div 5 = 4$

In the example on the left, when you subtract a number and then add the same number, the addition undoes the subtraction and you end up back at 184.

In the example on the right, you start with the number 4 and multiply it by 5 to get 20. And then you divide 20 by 5 to return to where you started at 4. So division is the inverse operation of multiplication.

Commutative operations

Addition and multiplication are both commutative operations. *Commutative* means that you can switch around the order of the numbers without changing the result. This property of addition and multiplication is called the *commutative property.* For example,

$$3 + 5 = 8 \text{ is the same as } 5 + 3 = 8$$

$$2 \cdot 7 = 14 \text{ is the same as } 7 \cdot 2 = 14$$

In contrast, subtraction and division are *noncommutative* operations. When you switch around the order of the numbers, the result changes. For example,

$$6 - 4 = 2, \text{ but } 4 - 6 = -2$$

$$5 \div 2 = \frac{5}{2} \text{ but } 2 \div 5 = \frac{2}{5}$$

Associative operations

Addition and multiplication are both *associative operations,* which means that you can group them differently without changing the result. This property of addition and multiplication is also called the *associative property.* Here's an example of how addition is associative. Suppose you want to add 3 + 6 + 2. You can solve this problem in two ways:

$(3 + 6) + 2$	$3 + (6 + 2)$
$= (9) + 2$	$= 3 + (8)$
$= 11$	$= 11$

And here's an example of how multiplication is associative. Suppose you want to multiply $5 \cdot 2 \cdot 4$. You can solve this problem in two ways:

$(5 \cdot 2) \cdot 4$	$5 \cdot (2 \cdot 4)$
$= 10 \cdot 4$	$= 5 \cdot 8$
$= 40$	$= 40$

In contrast, subtraction and division are *nonassociative* operations. This means that grouping them in different ways changes the result.

Distributing to lighten the load

In math, distribution (also called the *distributive property* of multiplication over addition) allows you to split a large multiplication problem into two smaller ones and add the results to get the answer.

For example, suppose you want to multiply $17 \cdot 101$. You can multiply them out, but distribution provides a different way to think about the problem that you may find easier. Because $101 = 100 + 1$, you can split this problem into two easier problems as follows:

$$= 17 \cdot (100 + 1)$$

$$= (17 \cdot 100) + (17 \cdot 1)$$

You take the number outside the parentheses, multiply it by each number inside the parentheses one at a time, then add the products. At this point, you may be able to solve the two multiplications in your head and then add them up easily:

$$= 1{,}700 + 17 = 1{,}717$$

Other Operations: Exponents, Square Roots, and Absolute Values

In this section, I introduce you to three new operations that you need as you move on with math: exponents, square roots, and absolute values. As with the Big Four operations, these three operations take numbers and tweak them in various ways.

Understanding exponents

Exponents (also called *powers*) are shorthand for repeated multiplication. For example, 2^3 means to multiply 2 by itself 3 times. To do that, use the following notation:

$$2^3 = 2 \cdot 2 \cdot 2 = 8$$

In this example, 2 is the *base number* and 3 is the *exponent.* You can read 2^3 as "two to the third power" or "two to the power of 3" (or even "two cubed," which has to do with the formula for finding the volume of a cube — see Chapter 11 for details).

When the base number is 10, figuring out any exponent is easy. Just write down a 1 and that many 0s after it:

$10^2 = 100$ (1 with two 0s)

$10^7 = 10{,}000{,}000$ (1 with seven 0s)

$10^{20} = 100{,}000{,}000{,}000{,}000{,}000{,}000$ (1 with twenty 0s)

The most common exponent is the number 2. When you take any whole number to the power of 2, the result is a square number. For this reason, taking a number to the power of 2 is called *squaring* that number. You can read 3^2 as "three squared," 4^2 as "four squared," and so forth.

Any number raised to the 0 power equals 1. So 1^0, 37^0, and $999{,}999^0$ are equivalent, or equal.

Discovering your roots

Earlier in this chapter, in "Fun and Useful Properties of the Big Four Operations," I show you how addition and subtraction are inverse operations. I also show you how multiplication and division are inverse operations. In a similar way, roots are the inverse operation of exponents.

The most common root is the square root. A *square root* undoes an exponent of 2. For example,

$$4^2 = 4 \cdot 4 = 16, \text{ so } \sqrt{16} = 4$$

You can read the symbol $\sqrt{}$ either as "the square root of" or as "radical." So read $\sqrt{16}$ as either "the square root of 16" or "radical 16."

You probably won't use square roots too much until you get to algebra, but at that point they become very handy.

Figuring out absolute value

The *absolute value* of a number is the positive value of that number. It tells you how far away from 0 a number is on the number line. The symbol for absolute value is a set of vertical bars.

Taking the absolute value of a positive number doesn't change that number's value. For example,

$|12| = 12$

$|145| = 145$

However, taking the absolute value of a negative number changes it to a positive number:

$|-5| = 5$

$|-212| = 212$

Finding Factors

In this section, I show you the relationship between factors and multiples. When one number is a *factor* of a second number, the second number is a *multiple* of the first number. For example, 20 is divisible by 5, so 5 is a factor of 20 and 20 is a multiple of 5.

Generating factors

You can easily tell whether a number is a factor of a second number: Just divide the second number by the first. If it divides evenly (with no remainder), the number is a factor; otherwise, it's not a factor.

For example, suppose you want to know whether 7 is a factor of 56. Because 7 divides 56 without leaving a remainder, 7 is a factor of 56. This method works no matter how large the numbers are.

The *greatest factor* of any number is the number itself, so you can always list all the factors of any number because you have a stopping point. Here's how to list all the factors of a number:

1. **Begin the list with 1, leave some space for other numbers, and end the list with the number itself.**

   Suppose you want to list all the factors of the number 18. Following these steps, you begin your list with 1 and end it with 18.

2. **Test whether 2 is a factor — that is, see whether the number is divisible by 2.**

   If it is, add 2 to the list, along with the original number divided by 2 as the second-to-last number on the list. For instance, 18 ÷ 2 = 9, so add 2 and 9 to the list of factors of 18.

3. **Test the number 3 in the same way.**

   You see that 18 ÷ 3 = 6, so add 3 and 6 to the list.

4. **Continue testing numbers until the beginning of the list meets the end of the list.**

   Check every number between to see whether it's evenly divisible. If it is, that number is also a factor. You get remainders when you divide 18 by 4 or 5, so the complete list of factors of 18 is 1, 2, 3, 6, 9, and 18.

A *prime number* is divisible only by 1 and itself — for example, the number 7 is divisible only by 1 and 7. On the other hand, a *composite number* is divisible by at least one number other than 1 and itself — for example, the number 9 is divisible not only by 1 and 9 but also by 3. A number's *prime factors* are the set of prime numbers (including repeats) that equal that number when multiplied together.

Finding the greatest common factor (GCF)

The *greatest common factor* (GCF) of a set of numbers is the largest number that's a factor of all those numbers. For example, the GCF of the numbers 4 and 6 is 2, because 2 is the greatest number that's a factor of both 4 and 6.

To find the GCF of a set of numbers, list all the factors of each number, as I show you in "Generating factors." The greatest factor appearing on every list is the GCF.

For example, suppose you want to find the GCF of 28, 42, and 70. Start by listing the factors of each:

- **Factors of 28:** 1, 2, 4, 7, 14, 28
- **Factors of 42:** 1, 2, 3, 6, 7, 14, 21, 42
- **Factors of 70:** 1, 2, 5, 7, 10, 14, 35, 70

The largest factor that appears on all three lists is 14; thus, the GCF of 28, 42, and 70 is 14.

Finding Multiples

Even though multiples tend to be larger numbers than factors, most students find them easier to work with. Read on for info on finding multiples and identifying the least common multiple of a set of numbers.

Generating multiples

The earlier section "Finding Factors" tells you how to find *all* the factors of a number. Finding all the factors is possible because a number's factors are always less than or equal to the number itself. So no matter how large a number is, it always has a *finite* (limited) number of factors.

Unlike factors, multiples of a number are greater than or equal to the number itself. (The only exception to this is 0, which is a multiple of every number.) Because of this, the multiples of a number go on forever — that is, they're *infinite.* Nevertheless, generating a partial list of multiples for any number is simple.

To list multiples of any number, write down that number and then multiply it by 2, 3, 4, and so forth.

For example, here are the first few positive multiples of 7:

7 14 21 28 35 42

As you can see, this list of multiples is simply part of the multiplication table for the number 7.

Finding the least common multiple (LCM)

The *least common multiple* (LCM) of a set of numbers is the lowest positive number that's a multiple of every number in that set.

To find the LCM of a set of numbers, take each number in the set and jot down a list of the first several multiples in order. The LCM is the first number that appears on every list.

When looking for the LCM of two numbers, start by listing multiples of the higher number, but stop this list when the number of multiples you've written down equals the lower number. Then start listing multiples of the lower number until one of them matches a number in the first list.

For example, suppose you want to find the LCM of 4 and 6. Begin by listing multiples of the higher number, which is 6. In this case, list only four of these multiples, because the lower number is 4.

Multiples of 6: 6, 12, 18, 24, ...

Now start listing multiples of 4:

Multiples of 4: 4, 8, 12, ...

Because 12 is the first number to appear on both lists of multiples, 12 is the LCM of 4 and 6.

Chapter 2

Evaluating Arithmetic Expressions

In This Chapter

- Understanding equations, expressions, and evaluation
- Doing the Big Four operations in the right order
- Working with expressions that contain exponents
- Evaluating expressions with parentheses

In this chapter, I introduce you to what I call the Three E's of math: equations, expressions, and evaluation.

You probably already know that an *equation* is a mathematical statement that has an equal sign (=) — for example, $1 + 1 = 2$. An *expression* is a string of mathematical symbols that you can place on one side of an equation — for example, $1 + 1$. And *evaluation* is finding out the *value* of an expression as a number — for example, finding out that the expression $1 + 1$ is equal to the number 2.

Throughout the rest of the chapter, I show you how to turn expressions into numbers using a set of rules called the *order of operations* (or *order of precedence*). These rules look complicated, but I break them down so you can see for yourself what to do next in any situation.

The Three E's: Equations, Expressions, and Evaluations

You should find the Three E's of math very familiar because whether you realize it or not, you've been using them for a long time. Whenever you add up the cost of several items at the store, balance your checkbook, or figure out the area of your room, you're evaluating expressions and setting up equations. In this section, I shed light on this stuff and give you a new way to look at it.

Equality for all: Equations

An *equation* is a mathematical statement that tells you that two things have the same value — in other words, it's a statement with an equal sign. The equation is one of the most important concepts in mathematics because it allows you to boil down a bunch of complicated information into a single number.

Mathematical equations come in lots of varieties: arithmetic equations, algebraic equations, differential equations, partial differential equations, Diophantine equations, and many more. In this book, you look at only two types: arithmetic equations and algebraic equations.

In this chapter, I discuss only *arithmetic equations,* which are equations involving numbers, the Big Four operations, and the other basic operations I introduce in Chapter 1 (absolute values, exponents, and roots). In Chapter 9, I introduce you to algebraic equations. Here are a few examples of simple arithmetic equations:

$$2 + 2 = 4$$

$$3 \cdot 4 = 12$$

$$20 \div 2 = 10$$

And here are a few examples of more-complicated arithmetic equations:

$$1{,}000 - 1 - 1 - 1 = 997$$

$$(1 \cdot 1) + (2 \cdot 2) = 5$$

$$4^2 - \sqrt{256} = (791 - 842) \cdot 0$$

Hey, it's just an expression

An *expression* is any string of mathematical symbols that can be placed on one side of an equation. Mathematical expressions, just like equations, come in a lot of varieties. In this chapter, I focus only on *arithmetic expressions,* which are expressions that contain numbers, the Big Four operations, and a few other basic operations (see Chapter 1). In Chapter 8, I introduce you to algebraic expressions.

Here are a few examples of simple expressions:

$$2 + 2$$

$$-17 + (-1)$$

$$14 \div 7$$

And here are a few examples of more-complicated expressions:

$$(88 - 23) \div 13$$

$$100 + 2 - 3 \cdot 17$$

$$\sqrt{441} + \left|-2^3\right|$$

Evaluating the situation

At the root of the word *evaluation* is the word *value.* When you evaluate something, you find its value. Evaluating an expression is also referred to as *simplifying, solving,* or *finding the value of* an expression. The words may change, but the idea is the same: boiling a string of numbers and math symbols down to a single number.

When you evaluate an arithmetic expression, you simplify it to a single numerical value — that is, you find the number that it's equal to. For example, evaluate the following arithmetic expression:

$$7 \cdot 5$$

How? Simplify it to a single number:

$$35$$

Putting the Three E's together

I'm sure you're dying to know how the Three E's — equations, expressions, and evaluation — are all connected. *Evaluation* allows you to take an *expression* containing more than one number and reduce it down to a single number. Then, you can make an *equation,* using an equal sign to connect the expression and the number. For example, here's an *expression* containing four numbers:

$$1 + 2 + 3 + 4$$

When you *evaluate* it, you reduce it down to a single number:

$$10$$

And now you can make an *equation* by connecting the expression and the number with an equal sign:

$$1 + 2 + 3 + 4 = 10$$

Following the Order of Operations

When you were a kid, did you ever try putting on your shoes first and then your socks? If you did, you probably discovered this simple rule:

1. **Put on socks.**
2. **Put on shoes.**

Thus, you have an order of operations: The socks have to go on your feet before your shoes. So in the act of putting on your shoes and socks, your socks have precedence over your shoes. A simple rule to follow, right?

In this section, I outline a similar set of rules for evaluating expressions called the *order of operations* (sometimes called *order of precedence*). Don't let the long name throw you. Order of operations is just a set of rules to make sure you get your socks and shoes on in the right order, mathematically speaking, so you always get the right answer.

Evaluate arithmetic expressions from left to right according to the following order of operations:

1. **Parentheses**
2. **Exponents**
3. **Multiplication and division**
4. **Addition and subtraction**

Don't worry about memorizing this list right now. I break it to you slowly in the remaining sections of this chapter, starting from the bottom and working toward the top, as follows:

- In "Order of operations and the Big Four expressions," I show Steps 3 and 4 — how to evaluate expressions with any combination of addition, subtraction, multiplication, and division.
- In "Order of operations in expressions with exponents," I show you how Step 2 fits in — how to evaluate expressions with Big Four operations *plus* exponents, square roots, and absolute values.
- In "Order of operations in expressions with parentheses," I show you how Step 1 fits in — how to evaluate all the expressions I explain *plus* expressions with parentheses.

Order of operations and the Big Four expressions

As I explain earlier in this chapter, evaluating an expression is just simplifying it down to a single number. Now I get you started on the basics of evaluating expressions that contain any combination of the Big Four operations — adding, subtracting, multiplying, and dividing. (For more on the Big Four, see Chapter 1.) Generally speaking, the Big Four expressions come in the three types outlined in Table 2-1.

Table 2-1 Types of Big Four Expressions

Expression	*Example*	*Rule*
Contains only addition and subtraction	$12 + 7 - 6 - 3 + 8$	Evaluate left to right.
Contains only multiplication and division	$18 \div 3 \cdot 7 \div 14$	Evaluate left to right.
Contains a combination of addition/subtraction and multiplication/division (mixed-operator expressions)	$9 + 6 \div 3$	1. Evaluate multiplication and division left to right. 2. Evaluate addition and subtraction left to right.

In this section, I show you how to identify and evaluate all three types of expressions.

Expressions with only addition and subtraction

Some expressions contain only addition and subtraction. When this is the case, the rule for evaluating the expression is simple.

When an expression contains only addition and subtraction, evaluate it step by step from left to right. For example, suppose you want to evaluate this expression:

$$17 - 5 + 3 - 8$$

Because the only operations are addition and subtraction, you can evaluate from left to right, starting with $17 - 5$:

$$= 12 + 3 - 8$$

As you can see, the number 12 replaces $17 - 5$. Now the expression has three numbers rather than four. Next, evaluate $12 + 3$:

$$= 15 - 8$$

This breaks the expression down to two numbers, which you can evaluate easily:

$$= 7$$

So $17 - 5 + 3 - 8 = 7$.

Expressions with only multiplication and division

Some expressions contain only multiplication and division. When this is the case, the rule for evaluating the expression is pretty straightforward.

When an expression contains only multiplication and division, evaluate it step by step from left to right. Suppose you want to evaluate this expression:

$$9 \cdot 2 \div 6 \div 3 \cdot 2$$

Again, the expression contains only multiplication and division, so you can move from left to right, starting with $9 \cdot 2$:

$$= 18 \div 6 \div 3 \cdot 2$$
$$= 3 \div 3 \cdot 2$$
$$= 1 \cdot 2$$
$$= 2$$

Notice that the expression shrinks one number at a time until all that's left is 2. So $9 \cdot 2 \div 6 \div 3 \cdot 2 = 2$.

Here's another quick example:

$$-2 \cdot 6 \div -4$$

Even though this expression has some negative numbers, the only operations it contains are multiplication and division. So you can evaluate it in two steps from left to right:

$$= -12 \div -4$$
$$= 3$$

Thus, $-2 \cdot 6 \div -4 = 3$.

Mixed-operator expressions

Often, an expression contains

- At least one addition or subtraction operator
- At least one multiplication or division operator

I call these *mixed-operator expressions*. To evaluate them, you need some stronger medicine. Here's the rule you want to follow.

Evaluate mixed-operator expressions as follows:

1. **Evaluate the multiplication and division from left to right.**
2. **Evaluate the addition and subtraction from left to right.**

For example, suppose you want to evaluate the following expression:

$$5 + 3 \cdot 2 + 8 \div 4$$

As you can see, this expression contains addition, multiplication, and division, so it's a mixed-operator expression. To evaluate it, start out by underlining the multiplication and division in the expression:

$$5 + \underline{3 \cdot 2} + \underline{8 \div 4}$$

Now, evaluate what you underlined from left to right:

$$= 5 + 6 + \underline{8 \div 4}$$
$$= 5 + 6 + 2$$

At this point, you're left with an expression that contains only addition, so you can evaluate it from left to right:

$$= 11 + 2$$
$$= 13$$

Thus, $5 + 3 \cdot 2 + 8 \div 4 = 13$.

Order of operations in expressions with exponents

Here's what you need to know to evaluate expressions that have exponents (see Chapter 1 for info on exponents).

Evaluate exponents from left to right *before* you begin evaluating Big Four operations (adding, subtracting, multiplying, and dividing).

The trick here is to turn the expression into a Big Four expression and then use what I show you earlier in "Order of operations and the Big Four expressions." For example, suppose you want to evaluate the following:

$$3 + 5^2 - 6$$

First, evaluate the exponent:

$$= 3 + 25 - 6$$

At this point, the expression contains only addition and subtraction, so you can evaluate it from left to right in two steps:

$$= 28 - 6$$
$$= 22$$

So $3 + 5^2 - 6 = 22$.

Order of operations in expressions with parentheses

In math, parentheses — ( ) — are often used to group together parts of an expression. When you're evaluating expressions, here's what you need to know about parentheses.

To evaluate expressions that contain parentheses, do the following:

1. **Evaluate the contents of the parentheses, from the inside out.**
2. **Evaluate the rest of the expression.**

Big Four expressions with parentheses

Suppose you want to evaluate $(1 + 15 \div 5) + (3 - 6) \cdot 5$. This expression contains two sets of parentheses, so evaluate these from left to right. Notice that the first set of parentheses contains a mixed-operator expression, so evaluate this in two steps starting with the division:

$$= (1+3)+(3-6)\cdot 5$$
$$= 4+(3-6)\cdot 5$$

Now evaluate the contents of the second set of parentheses:

$$= 4 + -3 \cdot 5$$

Now you have a mixed-operator expression, so evaluate the multiplication $(-3 \cdot 5)$ first, which gives you the following:

$$= 4 + -15$$

Finally, evaluate the addition:

$$= -11$$

So $(1 + 15 \div 5) + (3 - 6) \cdot 5 = -11$.

Expressions with exponents and parentheses

Try out the following example, which includes both exponents and parentheses:

$$1 + (3 - 6^2 \div 9) \cdot 2^2$$

Start out by working *only* with what's inside the parentheses. The first thing to evaluate there is the exponent, 6^2:

$$= 1 + (3 - 36 \div 9) \cdot 2^2$$

Continue working inside the parentheses by evaluating the division $36 \div 9$:

$$= 1 + (3 - 4) \cdot 2^2$$

Now you can get rid of the parentheses altogether:

$$= 1 + -1 \cdot 2^2$$

At this point, what's left is an expression with an exponent. This expression takes three steps, starting with the exponent:

$$= 1 + -1 \cdot 4$$
$$= 1 + -4$$
$$= -3$$

So $1 + (3 - 6^2 \div 9) \cdot 2^2 = -3$.

Expressions with parentheses raised to an exponent

Sometimes, the entire contents of a set of parentheses are raised to an exponent. In this case, evaluate the contents of the parentheses *before* evaluating the exponent, as usual. Here's an example:

$$(7 - 5)^3$$

First, evaluate 7 – 5:

$$= 2^3$$

With the parentheses removed, you're ready to evaluate the exponent:

$$= 2 \cdot 2 \cdot 2 = 8$$

Once in a rare while, the exponent itself contains parentheses. As always, evaluate what's in the parentheses first. For example,

$$21^{(19 + \underline{3 \cdot -6})}$$

This time, the smaller expression inside the parentheses is a mixed-operator expression. I underlined the part that you need to evaluate first:

$$21^{(19 + -18)}$$

Now you can finish off what's inside the parentheses:

$$= 21^1$$

At this point, all that's left is a very simple exponent:

$$= 21$$

So $21^{(19 + 3 \cdot -6)} = 21$.

Note: Technically, you don't need to put parentheses around the exponent. If you see an expression in the exponent, treat it as though it had parentheses around it. In other words, $21^{19 + 3 \cdot -6}$ means the same thing as $21^{(19 + 3 \cdot -6)}$.

Expressions with nested parentheses

Occasionally, an expression has *nested parentheses:* one or more sets of parentheses inside another set. Here, I give you the rule for handling nested parentheses.

When evaluating an expression with nested parentheses, evaluate what's inside the *innermost* set of parentheses first and work your way toward the *outermost* parentheses.

For example, suppose you want to evaluate the following expression:

$$2 + (9 - (\underline{7 - 3}))$$

I underlined the contents of the innermost set of parentheses, so evaluate these contents first:

$$= 2 + (9 - 4)$$

Next, evaluate what's inside the remaining set of parentheses:

$$= 2 + 5$$

Now you can finish things off easily:

$$= 7$$

So $2 + (9 - (7 - 3)) = 7$.

As a final example, here's an expression that requires everything from this chapter:

$$4 + (-7 \cdot (2^{(\underline{5 - 1})} - 4 \cdot 6))$$

This expression is about as complicated as you're ever likely to see in pre-algebra: one set of parentheses containing another set, which contains a third set. To start you off, I underlined what's deep inside this third set of parentheses. This is where you begin evaluating:

$$= 4 + (-7 \cdot (\underline{2^4 - 4 \cdot 6}))$$

Now, what's left is one set of parentheses inside another set. Again, work from the inside out. The smaller expression here is $2^4 - 4 \cdot 6$, so evaluate the exponent first, then the multiplication, and finally the subtraction:

$$= 4 + \left(-7 \cdot \left(\underline{16 - 4 \cdot 6}\right)\right)$$
$$= 4 + \left(-7 \cdot \left(\underline{16 - 24}\right)\right)$$
$$= 4 + \left(-7 \cdot -8\right)$$

Only one more set of parentheses to go:

$$= 4 + 56$$

At this point, finishing up is easy:

$$= 60$$

Therefore, $4 + (-7 \cdot (2^{(5-1)} - 4 \cdot 6)) = 60$.

As I say earlier in this section, this problem is about as hard as they come at this stage of math. Copy it down and try solving it step by step with the book closed.

Chapter 3

Say What? Making Sense of Word Problems

In This Chapter

- Knowing the four steps to solving a word problem
- Jotting down simple word equations to condense important info
- Plugging numbers into the word equations to solve the problem
- Attacking more-complex word problems with confidence

The very mention of word problems — or *story problems,* as they're sometimes called — is enough to send a cold shiver of terror into the bones of the average math student. Many would rather swim across a moat full of hungry crocodiles than "figure out how many bushels of corn Farmer Brown picked" or "help Aunt Sylvia decide how many cookies to bake." But word problems help you understand the logic behind setting up equations in real-life situations, making math actually useful — even if the scenarios in the word problems you practice on are pretty far-fetched.

In this chapter, I show you how to solve a word problem in four simple steps. After you understand the basics, I show you how to solve more-complex problems. Some of these problems have longer numbers to calculate, and others may have more-complicated stories. In either case, you can see how to work through them step by step.

Handling Basic Word Problems

Generally speaking, solving a word problem involves four steps:

1. **Read through the problem and set up *word equations* — that is, equations that contain words as well as numbers.**
2. **Plug in numbers in place of words wherever possible to set up a regular math equation.**
3. **Use math to solve the equation.**
4. **Answer the question that the problem asks.**

Most of this book is about Step 3. However, this chapter and Chapters 7 and 11 are all about Steps 1 and 2. I show you how to break down a word problem sentence by sentence, jot down the information you need to solve the problem, and then substitute numbers for words to set up an equation.

When you know how to turn a word problem into an equation, the hard part is done. Then you can use the rest of what you find in this book to figure out how to do Step 3 — solve the equation. From there, Step 4 is usually pretty easy, though at the end of each example, I make sure you understand how to do it.

Turning word problems into word equations

The first step to solving a word problem is reading it and putting the information you find into a useful form. In this section, I show you how to squeeze the juice out of a word problem and leave the pits behind!

Jotting down information as word equations

Most word problems give you information about numbers, telling you exactly how much, how many, how fast, how big, and so forth. Here are some examples:

Nunu is spinning 17 plates.

The width of the house is 80 feet.

If the local train is going 25 miles per hour . . .

You need this information to solve the problem. And paper is cheap, so don't be afraid to use it. (If you're concerned about trees, write on the back of all that junk mail you get.) Have a piece of scrap paper handy and jot down a few notes as you read through a word problem.

For example, here's how you can jot down "Nunu is spinning 17 plates":

Nunu = 17

Here's how to note that "the width of the house is 80 feet":

Width = 80

The third example tells you, "If the local train is going 25 miles per hour. . . ." So you can jot down the following:

Local = 25

Don't let the word *if* confuse you. When a problem says "If so-and-so is true . . ." and then asks you a question, assume it *is* true and use this information to answer the question.

When you jot down information this way, you're really turning words into a more useful form called a *word equation.* A word equation has an equal sign like a math equation, but it contains both words and numbers.

Turning more-complex statements into word equations

When you start doing word problems, you notice that certain words and phrases show up over and over again. For example,

Bobo is spinning five fewer plates than Nunu.

The height of a house is half as long as its width.

The express train is moving three times faster than the local train.

You've probably seen statements such as these in word problems since you were first doing math. Statements like these look like English, but they're really math, so spotting them is important. You can represent each of these types of statements as word equations that also use Big Four operations. Look again at the first example:

> Bobo is spinning five fewer plates than Nunu.

You don't know the number of plates that either Bobo or Nunu is spinning. But you know that these two numbers are related. You can express this relationship like this:

> Bobo = Nunu – 5

This word equation is shorter than the statement it came from. And as you see in the next section, word equations are easy to turn into the math that you need to solve the problem.

Here's another example:

> The height of a house is half as long as its width.

You don't know the width or height of the house, but you know that these numbers are connected. You can express this relationship between the width and height of the house as the following word equation:

> Height = width ÷ 2

With the same type of thinking, you can express "The express train is moving three times faster than the local train" as this word equation:

> Express = 3 · local

As you can see, each of the examples allows you to set up a word equation using one of the Big Four operations — adding, subtracting, multiplying, and dividing.

Figuring out what the problem's asking

The end of a word problem usually contains the question that you need to answer to solve the problem. You can use word equations to clarify this question so you know right from the start what you're looking for.

For example, you can write the question, “Altogether, how many plates are Bobo and Nunu spinning?” as

Bobo + Nunu = ?

You can write the question “How tall is the house?” as

Height = ?

Finally, you can rephrase the question “What’s the difference in speed between the express train and the local train?” in this way:

Express – Local = ?

Plugging in numbers for words

After you’ve written out a bunch of word equations, you have the facts you need in a form you can use. Now you can often solve the problem by plugging numbers from one word equation into another. In this section, I show you how to use the word equations you built in the preceding section to solve three problems.

Example: Send in the clowns

Some problems involve simple addition or subtraction. Here’s an example:

> Bobo is spinning five fewer plates than Nunu (Bobo dropped a few). Nunu is spinning 17 plates. Altogether, how many plates are Bobo and Nunu spinning?

Here’s what you have already, just from reading the problem:

Nunu = 17

Bobo = Nunu – 5

Plugging in the information gives you the following:

Bobo = ~~Nunu~~ 17 – 5 = 12

The problem wants you to find out how many plates the two clowns are spinning together. That is, you need to find out the following:

> Bobo + Nunu = ?

Just plug in the numbers, substituting 12 for Bobo and 17 for Nunu:

> ~~Bobo~~ 12 + ~~Nunu~~ 17 = 29

So Bobo and Nunu are spinning 29 plates.

Example: Our house in the middle of our street

At times, a problem may note relationships that require you to use multiplication or division. Here's an example:

> The height of a house is half as long as its width, and the width of the house is 80 feet. How tall is the house?

You already have a head start from what you determined earlier in "Turning more-complex statements into word equations":

> Width = 80
>
> Height = Width ÷ 2

You can plug in information as follows, substituting 80 for the word *width:*

> Height = ~~Width~~ 80 ÷ 2 = 40

So you know that the height of the house is 40 feet.

Solving More-Challenging Word Problems

The skills I show you previously in "Handling Basic Word Problems" are important for solving any word problem because they streamline the process and make it simpler.

And what's more, you can use those same skills to find your way through more-complex problems. Problems become more-complex when

- The calculations become harder. (For example, instead of a dress costing $30, now it costs $29.95.)
- The amount of information in the problem increases. (For example, instead of two clowns, now you have five.)

Don't let problems like these scare you. In this section, I show you how to use your new problem-solving skills to solve more-difficult word problems.

When numbers get serious

A lot of problems that look tough aren't much more difficult than the problems I show you in the previous sections. For example, consider this problem:

> Aunt Effie has $732.84 hidden in her pillowcase, and Aunt Jezebel has $234.19 less than Aunt Effie has. How much money do the two women have altogether?

Even though the numbers are larger, the principle is still the same as in problems in the earlier sections. Start reading from the beginning: "Aunt Effie has $732.84." This text is just information to jot down as a simple word equation:

Effie = $732.84

Continuing, you read: "Aunt Jezebel has $234.19 less than Aunt Effie has." It's another statement you can write as a word equation:

Jezebel = Effie – $234.19

Now you can plug in the number $732.84 where you see Aunt Effie's name in the equation:

Jezebel = ~~Effie~~ $732.84 – $234.19

So far, the big numbers haven't been any trouble. At this point, though, you probably need to stop to do the subtraction:

$$\begin{array}{r} \$732.84 \\ -\$234.19 \\ \hline \$498.65 \end{array}$$

Now you can jot this information down as always:

Jezebel = $498.65

The question at the end of the problem asks you to find out how much money the two women have altogether. Here's how to represent this question as an equation:

Effie + Jezebel = ?

You can plug information into this equation:

~~Effie~~ $732.84 + ~~Jezebel~~ $498.65 = ?

Again, because the numbers are large, you probably have to stop to do the math:

$$\begin{array}{r} \$732.84 \\ +\$498.65 \\ \hline \$1{,}231.49 \end{array}$$

So altogether, Aunt Effie and Aunt Jezebel have $1,231.49.

As you can see, the procedure for solving this problem is basically the same as for the simpler problems in the earlier sections. The only difference is that you have to stop to do some addition and subtraction.

Lots of information

When the going gets tough, knowing the system for writing word equations really becomes helpful. Here's a word problem that's designed to scare you off — but with your new skills, you're ready for it:

> Four women collected money to save the endangered Salt Creek tiger beetle. Keisha collected $160, Brie collected $50 more than Keisha, Amy collected twice as much as Brie, and together, Amy and Sophia collected $700. How much money did the four women collect altogether?

If you try to do this problem all in your head, you'll probably get confused. Instead, take it line by line and just jot down word equations as I discuss earlier in this chapter.

First, "Keisha collected $160." So jot down the following:

Keisha = 160

Next, "Brie collected $50 dollars more than Keisha," so write

Brie = Keisha + 50

After that, "Amy collected twice as much as Brie":

Amy = Brie · 2

And finally, "together, Amy and Sophia collected $700":

Amy + Sophia = 700

That's all the information that the problem gives you, so now you can start working with it. Keisha collected $160, so you can plug in 160 anywhere you find Keisha's name:

Brie = ~~Keisha~~ 160 + 50 = 210

Now you know how much Brie collected, so you can plug this information into the next equation:

Amy = ~~Brie~~ 210 · 2 = 420

This equation tells you how much Amy collected, so you can plug this number into the last equation:

~~Amy~~ 420 + Sophia = 700

To solve this problem, change it from addition to subtraction using inverse operations, as I show you in Chapter 1:

Sophia = 700 – 420 = 280

Now that you know how much money each woman collected, you can answer the question at the end of the problem:

Keisha + Brie + Amy + Sophia = ?

You can plug in this information easily:

~~Keisha~~ 160 + ~~Brie~~ 210 + ~~Amy~~ 420 + ~~Sophia~~ 280 = 1,070

So you can conclude that the four women collected $1,070 altogether.

Putting it all together

Here's one final example putting together everything from this chapter. Try writing down this problem and working it through step by step on your own. If you get stuck, come back here. When you can solve it from beginning to end with the book closed, you'll have a good grasp of how to solve word problems:

> On a recent shopping trip, Travis bought six shirts for $19.95 each and two pairs of pants for $34.60 each. He then bought a jacket that cost $37.08 less than he paid for both pairs of pants. If he paid the cashier with three $100 bills, how much change did he receive?

On the first read-through, you may wonder how Travis found a store that prices jackets that way. (Believe me — it was quite a challenge.) Anyway, back to the problem. You can jot down the following word equations:

Shirts = $19.95 · 6

Pants = $34.60 · 2

Jacket = Pants – $37.08

The numbers in this problem are probably longer than you can solve in your head, so they require some attention:

$$\begin{array}{r} \$19.95 \\ \times \quad 6 \\ \hline \$119.70 \end{array} \qquad \begin{array}{r} \$34.60 \\ \times \quad 2 \\ \hline \$69.20 \end{array}$$

With this done, you can fill in some more information:

Shirts = $119.70

Pants = $69.20

Jacket = Pants – $37.08

Now you can plug in $69.20 for *pants* to find the cost of the jacket:

Jacket = ~~Pants~~ $69.20 – $37.08

Again, because the numbers are long, solve this equation separately:

$$\begin{array}{r} \$69.20 \\ -\$37.08 \\ \hline \$32.12 \end{array}$$

This equation gives you the price of the jacket:

Jacket = $32.12

Now that you have the price of the shirts, pants, and jacket, you can find out how much Travis spent:

Amount Travis spent = ~~Shirts~~ $119.70 + ~~Pants~~ $69.20 + ~~Jacket~~ $32.12

Again, you have another equation to solve:

$$\begin{array}{r} \$119.70 \\ \$69.20 \\ +\$32.12 \\ \hline \$221.02 \end{array}$$

So you can jot down the following:

Amount Travis spent = $221.02

The problem is asking you to find out how much change Travis received from $300, so jot this down:

Change = $300 – Amount Travis spent

You can plug in the amount that Travis spent:

Change = \$300 – \$221.02

And do just one more equation:

$$\begin{array}{r} \$300.00 \\ \underline{-\$221.02} \\ \$78.98 \end{array}$$

So you can jot down the answer:

Change = \$78.98

Therefore, Travis received \$78.98 in change.

Chapter 4

Figuring Out Fractions

In This Chapter

- Looking at multiplication and division of fractions
- Adding and subtracting fractions in a bunch of different ways
- Applying the Big Four operations to mixed numbers

Fractions have a few important properties that are worth knowing right from the start. For instance, when the *numerator* (top number) and the *denominator* (bottom number) are equal, the fraction equals 1. When the numerator is less than the denominator, the fraction is less than 1. Fractions like these are called *proper fractions.* Positive proper fractions are always between 0 and 1.

However, when the numerator is greater than the denominator, the fraction is *greater than* 1. Any fraction that's greater than 1 is called an *improper fraction.* It's customary to convert an improper fraction to a mixed number, especially when it's the final answer to a problem. A *mixed number* is a combination of a whole number and a proper fraction added together.

In this chapter, the focus is on applying the Big Four operations to fractions. I start out by showing you how to multiply and divide fractions, which isn't much more difficult than multiplying whole numbers. Surprisingly, adding and subtracting fractions is a bit trickier.

Later in the chapter, I move on to mixed numbers. Again, multiplication and division shouldn't pose too much of a problem because the process in each case is almost the same as multiplying and dividing fractions. I save adding and subtracting mixed numbers for the very end. By then, you should be much more comfortable with fractions and ready to tackle the challenge.

Reducing Fractions to Lowest Terms

Reducing a fraction to its lowest terms involves identifying a common factor in the numerator and denominator. A fraction is fully reduced, or in its lowest terms, when the numerator and denominator do not share any common factors.

Here's a fraction that's not in its lowest terms:

$$\frac{28}{40}$$

Write down the factors of both the numerator and denominator.

Factors of 28: 1, 2, 4, 7, 28

Factors of 40: 1, 2, 4, 5, 8, 10, 20, 40

Find the *greatest common factor,* or the largest factor found in both lists. The largest factor of both numbers is 4.

Divide the numerator and denominator by the greatest common factor.

$$\frac{28 \div 4}{40 \div 4} = \frac{7}{10}$$

The numerator and denominator do not share any other common factors, so this fraction is now fully reduced.

Multiplying and Dividing Fractions

One of the odd little ironies of life is that multiplying and dividing fractions is easier than adding or subtracting them. For this reason, I show you how to multiply and divide fractions before I show you how to add or subtract them.

Multiplying numerators and denominators straight across

Everything in life should be as simple as multiplying fractions. All you need for multiplying fractions is a pen or pencil, something to write on, and a basic knowledge of the multiplication table.

Here's how to multiply two fractions:

1. **Multiply the *numerators* (the numbers on top) together to get the numerator of the answer.**
2. **Multiply the *denominators* (the numbers on the bottom) together to get the denominator of the answer.**

For example, here's how to multiply $\frac{2}{5} \cdot \frac{3}{7}$:

$$\frac{2}{5} \cdot \frac{3}{7} = \frac{2 \cdot 3}{5 \cdot 7} = \frac{6}{35}$$

Sometimes, when you multiply fractions, you may have an opportunity to reduce to lowest terms (see the preceding section for details).

Doing a flip to divide fractions

Dividing fractions is just as easy as multiplying them. In fact, when you divide fractions, you really turn the problem into multiplication.

To divide one fraction by another, multiply the first fraction by the reciprocal of the second. The *reciprocal* of a fraction is simply that fraction turned upside down.

For example, here's how you turn fraction division into multiplication:

$$\frac{1}{3} \div \frac{4}{5} = \frac{1}{3} \cdot \mathbf{\frac{5}{4}}$$

As you can see, I turn $\frac{4}{5}$ into its reciprocal — $\frac{5}{4}$ — and change the division sign to a multiplication sign. After that, just multiply the fractions as I describe in "Multiplying numerators and denominators straight across":

$$\frac{1}{3}\cdot\frac{5}{4}=\frac{1\cdot 5}{3\cdot 4}=\frac{5}{12}$$

Adding Fractions

When you add fractions, one important thing to notice is whether their denominators (the numbers on the bottom) are the same. If they're the same, adding fractions that have the same denominator is a walk in the park. But when fractions have different denominators, adding them becomes a bit more complex.

In this section, I first show you how to add fractions with the same denominator. Then I show you a method for adding fractions when the denominators are different.

Finding the sum of fractions with the same denominator

To add two fractions that have the same denominator (bottom number), add the numerators (top numbers) together and leave the denominator unchanged.

For example, consider the following problem:

$$\frac{1}{5}+\frac{2}{5}=\frac{1+2}{5}=\frac{3}{5}$$

As you can see, to add these two fractions, you add the numerators (1 + 2) and keep the denominator (5) the same.

Even if you have to add more than two fractions, as long as the denominators are all the same, you just add the numerators and leave the denominator unchanged:

$$\frac{1}{17}+\frac{3}{17}+\frac{4}{17}+\frac{6}{17}=\frac{1+3+4+6}{17}=\frac{14}{17}$$

Sometimes, when you add fractions with the same denominator, you may have to reduce it to lowest terms. Take this problem for example:

$$\frac{1}{4}+\frac{1}{4}=\frac{1+1}{4}=\frac{2}{4}$$

The numerator and the denominator are both even, so you know they can be reduced:

$$\frac{2}{4}=\frac{1}{2}$$

In other cases, the sum of two proper fractions is an improper fraction. You get a numerator that's larger than the denominator when the two fractions add up to more than 1, as in this case:

$$\frac{3}{7}+\frac{5}{7}=\frac{8}{7}$$

If you have more work to do with this fraction, leave it as an improper fraction so that it's easier to work with. But if this is your final answer, you may need to turn it into a mixed number, as I show you later in "Converting between improper fractions and mixed numbers":

$$\frac{8}{7}=8\div 7=1\text{ r }1=1\frac{1}{7}$$

When two fractions have the same *numerator* (top number), don't add them by adding the denominators and leaving the numerator unchanged.

Adding fractions with different denominators

When the fractions that you want to add have different denominators, adding them isn't quite as easy.

Here is one method for adding fractions with different denominators:

1. **Cross-multiply the two fractions and add the results together to get the numerator of the answer.**

Suppose you want to add the fractions $\frac{1}{3}$ and $\frac{2}{5}$. To get the numerator of the answer, *cross-multiply.* In other words, multiply the numerator of each fraction by the denominator of the other fraction:

$$\frac{1}{3}+\frac{2}{5}$$

$$1\cdot5=5$$

$$2\cdot3=6$$

Add the results to get the numerator of the answer:

$$5+6=11$$

2. **Multiply the two denominators together to get the denominator of the answer.**

   To get the denominator, just multiply the denominators (bottom numbers) of the two fractions:

   $$3\cdot5=15$$

   The denominator of the answer is 15.

3. **Write your answer as a fraction.**

   $$\frac{1}{3}+\frac{2}{5}=\frac{11}{15}$$

In some cases, you may have to add more than one fraction. The method is similar, with one small tweak. For example, suppose you want to add $\frac{1}{2}+\frac{3}{5}+\frac{4}{7}$:

1. **Start out by multiplying the *numerator* of the first fraction by the *denominators* of all the other fractions.**

   $$\frac{\mathbf{1}}{2}+\frac{3}{\mathbf{5}}+\frac{4}{\mathbf{7}}$$

   $$(1\cdot5\cdot7)=35$$

2. **Do the same with the second fraction and add this value to the first.**

   $$\frac{1}{\mathbf{2}}+\frac{\mathbf{3}}{5}+\frac{4}{\mathbf{7}}$$

   $$35+(3\cdot2\cdot7)=35+42$$

3. **Do the same with the remaining fraction(s).**

$$\frac{1}{2}+\frac{3}{5}+\frac{\mathbf{4}}{7}$$

$$35+42+(4\cdot 2\cdot 5)=35+42+40=117$$

When you're done, you have the numerator of the answer.

4. **To get the denominator, just multiply all the denominators together:**

$$\frac{1}{\mathbf{2}}+\frac{3}{\mathbf{5}}+\frac{4}{\mathbf{7}}$$

$$=\frac{35+42+40}{2\cdot 5\cdot 7}=\frac{117}{70}$$

As usual, you may need to reduce or change an improper fraction to a mixed number. In this example, you just need to change to a mixed number (as I explain in "Converting between improper fractions and mixed numbers"):

$$\frac{117}{70}=117\div 70=1\text{ r }47=1\frac{47}{70}$$

Subtracting Fractions

Subtracting fractions isn't really much different from adding them. As with addition, when the denominators are the same, subtraction is easy. And when the denominators are different, I show you a different approach, the *traditional method,* for subtracting fractions.

Subtracting fractions with the same denominator

When two fractions have the same denominator (bottom number), here's how you subtract one fraction from another: Subtract the numerator (top number) of the second fraction from the numerator of the first fraction and keep the denominator the same. For example,

$$\frac{3}{5}-\frac{2}{5}=\frac{3-2}{5}=\frac{1}{5}$$

Sometimes, as when you subtract fractions, you may have to reduce:

$$\frac{3}{10}-\frac{1}{10}=\frac{3-1}{10}=\frac{2}{10}$$

Because the numerator and denominator are both even, you can reduce this fraction by a factor of 2 (as I show you earlier in "Reducing Fractions to Lowest Terms"):

$$\frac{2}{10}=\frac{2\div 2}{10\div 2}=\frac{1}{5}$$

Unlike addition, when you subtract one proper fraction from another, you never get an improper fraction.

Subtracting fractions with different denominators

Similar to addition, you have a choice of methods when subtracting fractions. In this section, I show you the traditional way to convert fractions with two different denominators. You can use this method to add fractions as well.

To use the traditional way to subtract fractions with two different denominators, follow these steps:

1. **Find the least common multiple (LCM) of the two denominators (for more on finding the LCM of two numbers, see Chapter 1).**

   For example, suppose you want to subtract $\frac{7}{8}-\frac{11}{14}$. Here's how to find the LCM of 8 and 14. List the first eight multiples of 14, and then list multiples of 8 until you find a number that appears in both lists::

   Multiples of 14: 14, 28, 42, 56, 70, 84, 98, 112, ...

   Multiples of 8: 8, 16, 24, 32, 40, 48, 56, ...

   The LCM is 56.

2. **Increase each fraction to higher terms so that the denominator of each equals the LCM.**

The denominators of both should be 56:

$$\frac{7}{8} = \frac{7 \cdot 7}{8 \cdot 7} = \frac{49}{56}$$

$$\frac{11}{14} = \frac{11 \cdot 4}{14 \cdot 4} = \frac{44}{56}$$

3. **Substitute these two new fractions for the original ones and subtract as I show you earlier in "Subtracting fractions with the same denominator."**

$$\frac{49}{56} - \frac{44}{56} = \frac{5}{56}$$

This time, you don't need to reduce, because 5 is a prime number and 56 isn't divisible by 5. In some cases, however, you have to reduce the answer to lowest terms.

Working with Mixed Numbers

Both the methods I describe earlier in this chapter work for both proper and improper fractions. Unfortunately, mixed numbers are ornery little critters, and you need to figure out how to deal with them on their own terms.

Converting between improper fractions and mixed numbers

Improper fractions are often easier to work with, but answers usually need to be mixed numbers. This section shows you how to convert between mixed numbers and improper fractions.

Switching to an improper fraction

Here's how to convert a mixed number to an improper fraction:

1. **Multiply the denominator of the fractional part by the whole number, and add the result to the numerator.**

For example, suppose you want to convert the mixed number $5\frac{2}{3}$ to an improper fraction. First, multiply 3 by 5 and add 2:

$$(3 \cdot 5) + 2 = 17$$

2. **Use this result as your numerator, and place it over the denominator you already have.**

   Place 17 over the denominator:

   $$\frac{17}{3}$$

   So the mixed number $5\frac{2}{3}$ equals the improper fraction $\frac{17}{3}$.

Switching to a mixed number

To convert an improper fraction to a mixed number, divide the numerator by denominator. Then write the mixed number this way:

- The quotient (answer) is the whole-number part.
- The remainder is the numerator.
- The denominator of the improper fraction is the denominator.

For example, suppose you want to write the improper fraction $\frac{19}{5}$ as a mixed number. First, divide 19 by 5. The answer is 3 with a remainder of 4:

$$19 \div 5 = 3 \text{ r } 4$$

Then write the mixed number as follows:

$$3\frac{4}{5}$$

Multiplying and dividing mixed numbers

I can't give you a direct method for multiplying and dividing mixed numbers. The only way is to convert the mixed numbers to improper fractions and multiply or divide as usual.

Here's how to multiply or divide mixed numbers:

1. **Convert all mixed numbers to improper fractions.**

For example, suppose you want to multiply $1\frac{3}{5} \cdot 2\frac{1}{3}$. First convert $1\frac{3}{5}$ and $2\frac{1}{3}$ to improper fractions:

$$1\frac{3}{5} = \frac{5 \cdot 1 + 3}{5} = \frac{8}{5}$$

$$2\frac{1}{3} = \frac{3 \cdot 2 + 1}{3} = \frac{7}{3}$$

2. **Multiply these improper fractions.**

$$\frac{8}{5} \cdot \frac{7}{3} = \frac{8 \cdot 7}{5 \cdot 3} = \frac{56}{15}$$

3. **If the answer is an improper fraction, convert it back to a mixed number.**

$$\frac{56}{15} = 56 \div 15 = 3 \text{ r } 11 = 3\frac{11}{15}$$

In this case, the answer is already in lowest terms, so you don't have to reduce it.

Adding and subtracting mixed numbers

One way to add and subtract mixed numbers is to convert them to improper fractions and then to add or subtract them as usual. However, you can also work with the fractional and whole-number parts separately, as I show you in this section.

Adding two mixed numbers

Adding mixed numbers looks a lot like adding whole numbers. For this reason, some students feel more comfortable adding mixed numbers than adding fractions.

Here's how to add two mixed numbers:

1. **Add the fractional parts using any method you like, and if necessary, reduce the fraction.**
2. **If the answer you found in Step 1 is an improper fraction, change it to a mixed number, write down the fractional part, and carry the whole-number part to the whole-number column.**

3. **Add the whole-number parts (including any number carried).**

For example, suppose you want to add $8\frac{3}{5} + 6\frac{4}{5}$. Here's how you do it:

1. **Add the fractions.**

   $$\frac{3}{5} + \frac{4}{5} = \frac{7}{5}$$

2. **Switch improper fractions to mixed numbers, write down the fractional part, and carry over the whole number.**

   Because the sum is an improper fraction, convert it to the mixed number $1\frac{2}{5}$. Write down $\frac{2}{5}$ and carry the 1 over to the whole-number column.

3. **Add the whole-number parts, including any whole numbers you carried over when you switched to a mixed number.**

   $$1 + 8 + 6 = 15$$

Here's how the solved problem looks in column form. (Be sure to line up the whole numbers in one column and the fractions in another.)

$$\begin{array}{r} {}^{1}8\frac{3}{5} \\ +6\frac{4}{5} \\ \hline 15\frac{2}{5} \end{array}$$

You see the most difficult type of mixed-number addition when the denominators of the fractions are different. This difference doesn't change Steps 2 or 3, but it does make Step 1 tougher.

For example, suppose you want to add $16\frac{3}{5}$ and $7\frac{7}{9}$.

1. **Add the fractions.**

   Add $\frac{3}{5}$ and $\frac{7}{9}$. As I show you earlier in "Adding fractions with different denominators," you get the numerator of the answer by cross-multiplying the two fractions and adding the results (3 · 9 + 7 · 5); you get the new denominator by multiplying the two denominators (5 · 9):

$$\frac{3}{5}+\frac{7}{9}=\frac{3\cdot 9+7\cdot 5}{5\cdot 9}=\frac{27+35}{45}=\frac{62}{45}$$

2. **Switch improper fractions to mixed numbers, write down the fractional part, and carry over the whole number.**

   The fraction $\frac{62}{45}$ is improper, so change it to the mixed number $1\frac{17}{45}$. Fortunately, the fractional part of this mixed number isn't reducible. Write down the $\frac{17}{45}$ and carry over the 1 to the whole-number column.

3. **Add the whole numbers.**

   1 + 16 + 7 = 24

Here's how the completed problem looks:

$$\begin{array}{r} \overset{1}{16}\frac{3}{5} \\ +\ 7\frac{7}{9} \\ \hline 24\frac{17}{45} \end{array}$$

Subtracting mixed numbers

The basic way to subtract mixed numbers is close to the way you add them. Again, the subtraction looks more like what you're used to with whole numbers.

Here's how to subtract two mixed numbers:

1. **Find the difference of the fractional parts.**
2. **Find the difference of the two whole-number parts.**

As with addition, subtraction is much easier when the denominators are the same.

Borrowing with mixed numbers

One complication arises when you try to subtract a larger fractional part from a smaller one. Suppose you want to find $11\frac{1}{6} - 2\frac{5}{6}$. This time, if you try to subtract the fractions, you get

$$\frac{1}{6} - \frac{5}{6} = -\frac{4}{6}$$

Obviously, you don't want to end up with a negative number in your answer. You can handle this problem by borrowing from the column to the left. This idea is very similar to the borrowing that you use in regular subtraction, with one key difference.

When borrowing in mixed-number subtraction, do the following:

1. **Borrow 1 from the whole-number portion and add it to the fractional portion, turning the fraction into a mixed number.**

   To find $11\frac{1}{6} - 2\frac{5}{6}$, borrow 1 from the 11 and add it to $\frac{1}{6}$, making it the mixed number $1\frac{1}{6}$:

   $$11\frac{1}{6} = 10 + 1\frac{1}{6}$$

2. **Change this new mixed number into an improper fraction.**

   Here's what you get when you change $1\frac{1}{6}$ into an improper fraction:

   $$10 + 1\frac{1}{6} = 10\frac{7}{6}$$

   The result is $10\frac{7}{6}$. This answer is a weird cross between a mixed number and an improper fraction, but it's what you need to handle the job.

3. **Use the result in your subtraction.**

   $$\begin{array}{r} 10\frac{7}{6} \\ -\ 2\frac{5}{6} \\ \hline 8\frac{2}{6} \end{array}$$

In this case, you have to reduce the fractional part of the answer:

$$8\frac{2}{6} = 8\frac{1}{3}$$

Unequal denominators: Checking whether you need to borrow

Suppose you want to subtract $15\frac{4}{11} - 12\frac{3}{7}$. Because the denominators are different, subtracting the fractions becomes more difficult. But you have another question to think about: In this problem, do you need to borrow? If $\frac{4}{11}$ is greater than $\frac{3}{7}$, you don't have to borrow. But if $\frac{4}{11}$ is less than $\frac{3}{7}$, you do.

Here's how to test two fractions to see which is greater by cross-multiplying:

$$\begin{array}{cc} \frac{4}{11} & \frac{3}{7} \\ 4 \cdot 7 = 28 & 3 \cdot 11 = 33 \end{array}$$

Because 28 is less than 33, $\frac{4}{11}$ is less than $\frac{3}{7}$, so you do have to borrow. I get the borrowing out of the way first:

$$15\frac{4}{11} = 14 + 1\frac{4}{11} = 14\frac{15}{11}$$

Now the problem looks like this:

$$14\frac{15}{11} - 12\frac{3}{7}$$

The first step, subtracting the fractions, is going to be the most time-consuming, so as I show you earlier in "Subtracting fractions with different denominators," you can take care of that on the side:

$$\frac{15}{11} - \frac{3}{7} = \frac{15 \cdot 7 - 3 \cdot 11}{11 \cdot 7} = \frac{105 - 33}{77} = \frac{72}{77}$$

The good news is that this fraction can't be reduced. (It can't be reduced because 72 and 77 have no common factors: $72 = 2 \cdot 2 \cdot 2 \cdot 3 \cdot 3$, and $77 = 7 \cdot 11$.) So the hard part of the problem is done, and the rest follows easily:

$$\begin{array}{r} 14\frac{15}{11} \\ -\ 12\frac{3}{7} \\ \hline 2\frac{72}{77} \end{array}$$

This problem is about as difficult as a mixed-number subtraction problem gets. Look over it step by step. Or better yet, copy the problem and then close the book and try to work through the steps on your own. If you get stuck, that's okay. Better now than on an exam!

Chapter 5

Deciphering Decimals

In This Chapter

- Applying decimals to the Big Four operations
- Looking at decimal and fraction conversions
- Making sense of repeating decimals

Decimals look and feel more like whole numbers than fractions do, so when working with decimals, you don't have to worry about reducing and increasing terms, improper fractions, mixed numbers, and a lot of other stuff.

Performing the Big Four operations — addition, subtraction, multiplication, and division — on decimals is very close to performing them on whole numbers. The numerals 0 through 9 work just like they usually do. As long as you get the decimal point in the right place, you're home free.

In this chapter, I show you all about working with decimals. I also show you how to convert fractions to decimals and decimals to fractions. Finally, I give you a peek into the strange world of repeating decimals.

Performing the Big Four Operations with Decimals

Everything you already know about adding, subtracting, multiplying, and dividing whole numbers carries over when you work with decimals. In fact, in each case, there's really only one key difference: how to handle that pesky little decimal point. In this section, I show you how to perform the Big Four math operations with decimals.

Adding decimals

Adding decimals is almost as easy as adding whole numbers. As long as you set up the problem correctly, you're in good shape. To add decimals, follow these steps:

1. **Line up the decimal points.**
2. **Add as usual from right to left, column by column.**
3. **Place the decimal point in the answer in line with the other decimal points in the problem.**

For example, suppose you want to add the numbers 14.5 and 1.89. Line up the decimal points neatly as follows:

$$\begin{array}{r} 14.5 \\ +\ \ 1.89 \\ \hline \end{array}$$

Begin adding from the right-hand column. Treat the blank space after 14.5 as a 0 — you can write this in as a 0. Complete the problem column by column, and at the end, put the decimal point directly below the others in the problem:

$$\begin{array}{r} 1\overset{1}{4}.50 \\ +\ \ 1.89 \\ \hline 16.39 \end{array}$$

When adding more than one decimal, the same rules apply. For example, suppose you want to add 15.1 + 0.005 + 800 + 1.2345. The most important idea is lining up the decimal points correctly:

$$\begin{array}{r} 15.1 \\ 0.005 \\ 800.0 \\ +\ \ 1.2345 \\ \hline \end{array}$$

To avoid mistakes, be especially neat when adding a lot of decimals.

Because the number 800 isn't a decimal, I place a decimal point and a 0 at the end of it to be clear about how to line it up. If you like, you can make sure all numbers have the same number of decimal places (in this case, four) by adding trailing zeros. *Trailing zeros* are zeros written to the right of a decimal point and all nonzero digits, as in 15.1000. These extra zeros do not affect the value of a number. After you properly set up the problem, the addition is no more difficult than in any other addition problem:

$$\begin{array}{r} 15.1000 \\ 0.0050 \\ 800.0000 \\ +\ \ 1.2345 \\ \hline 816.3395 \end{array}$$

Subtracting decimals

Subtracting decimals uses the same trick as adding them (which I talk about in the preceding section). Here's how you subtract decimals:

1. **Line up the decimal points.**
2. **Subtract as usual from right to left, column by column.**
3. **When you're done, place the decimal point in the answer in line with the other decimal points in the problem.**

For example, suppose you want to figure out 144.87 – 0.321. First, line up the decimal points. In this case, I add a zero at the end of the first decimal. This placeholder reminds you that in the right-hand column, you need to borrow to get the answer to 0 – 1. The rest of the problem is very straightforward. Just finish the subtraction and drop the decimal point straight down:

$$\begin{array}{r} 1\,4\,4.\,8\,\overset{6}{\not{7}}10 \\ -\ \ 0.\,3\,2\,\ 1 \\ \hline 1\,4\,4.\,5\,4\,\ 9 \end{array}$$

As with addition, the decimal point in the answer goes directly below where it appears in the problem.

Multiplying decimals

Multiplying decimals is different from adding and subtracting them in that you don't have to worry about lining up the decimal points (see the preceding sections). In fact, the only difference between multiplying whole numbers and decimals comes at the very end.

Here's how to multiply decimals:

1. **Perform the multiplication as you would for whole numbers.**
2. **When you're done, count the number of digits to the right of the decimal point in each factor and add the result.**
3. **Place the decimal point in your answer so that your answer has the same number of digits after the decimal point.**

This process sounds tricky, but multiplying decimals can actually be simpler than adding or subtracting them. Suppose, for instance, you want to multiply 23.5 by 0.16. The first step is to pretend that you're multiplying numbers without decimal points.

Your answer isn't complete, though, because you still need to find out where the decimal point goes. To do this, notice that 23.5 has one digit after the decimal point and that 0.16 has two digits after the decimal point. Because $1 + 2 = 3$ digits, place the decimal point in the answer so that it has three digits after the decimal point. (You can put your pencil at the 0 at the end of 3760 and move the decimal point three places to the left.)

$$\begin{array}{r} 23.5 \\ \underline{\times 0.16} \\ 1410 \\ \underline{+\ 2350} \\ 3.760 \end{array}$$

Even though the last digit in the answer is a 0, you still need to count this as a digit when placing the decimal point. After the decimal point is in place, you can drop trailing zeros.

So the answer is 3.760, which is equal to 3.76.

Dividing decimals

Dividing decimals is almost the same as dividing whole numbers. The main difference comes at the beginning, before you start dividing.

Here's how to divide decimals:

1. **Turn the *divisor* (the number you're dividing by) into a whole number by moving the decimal point all the way to the right; at the same time, move the decimal point in the *dividend* (the number you're dividing) the same number of places to the right.**

   For example, suppose you want to divide 10.274 by 0.11. Write the problem as usual:

   $$0.11\overline{)10.274}$$

   Turn 0.11 into a whole number by moving the decimal point in 0.11 two places to the right, giving you 11. At the same time, move the decimal point in 10.274 two places to the right, giving you 1,027.4:

   $$11.\overline{)1027.4}$$

2. **Place a decimal point in the *quotient* (the answer) directly above where the decimal point now appears in the dividend.**

3. **Divide as usual, being careful to line up the quotient properly so that the decimal point falls into place.**

   $$\begin{array}{r} 93.4 \\ 11.\overline{)1027.4} \\ \underline{-99} \\ 37 \\ \underline{-33} \\ 44 \\ \underline{-44} \\ 0 \end{array}$$

   So the answer is 93.4. As you can see, as long as you're careful when placing the decimal point and the digits, the correct answer appears with the decimal point in the right position.

Dealing with more zeros in the dividend

Sometimes, you may have to add one or more trailing zeros to the dividend. You can add as many trailing zeros as you like to a decimal without changing its value. For example, suppose you want to divide 67.8 by 0.333:

1. **Change 0.333 into a whole number by moving the decimal point three places to the right; at the same time, move the decimal point in 67.8 three places to the right:**

   $333.\overline{)67800.}$

   In this case, when you move the decimal point in 67.8, you run out of room, so you have to add a couple of zeros to the dividend. This step is perfectly valid, and you need to do this whenever the divisor has more decimal places than the dividend.

2. **Place the decimal point in the quotient directly above where it appears in the dividend.**

3. **Divide as usual, being careful to line up the numbers in the quotient correctly.**

$$\begin{array}{r} 203. \\ 333.\overline{)67800.} \\ \underline{666} \\ 1200 \\ \underline{-999} \\ 201 \end{array}$$

This time, the division doesn't work out evenly. If this were a problem with whole numbers, you'd finish by writing down a remainder of 201. But decimals are a different story. The next section explains why, with decimals, the show must go on.

Completing decimal division

When you're dividing whole numbers, you can complete the problem simply by writing down the remainder. But remainders are *never* allowed in decimal division.

A common way to complete a problem in decimal division is to round off the answer. In most cases, you'll be instructed to round your answer to the nearest whole number or to one or two decimal places.

To complete a decimal division problem by rounding it off, you need to add at least one trailing zero to the dividend. Attaching a trailing zero doesn't change a decimal, but it does allow you to bring down one more number, changing 201 into 2,010. Now you can divide 333 into 2,010:

$$
\begin{array}{r}
203.6 \\
333.\overline{)67800.0} \\
\underline{666} \\
1200 \\
\underline{-999} \\
2010 \\
\underline{-1998} \\
12
\end{array}
$$

At this point, you can round the answer to the nearest whole number, 204.

Converting between Decimals and Fractions

Fractions (see Chapter 4) and decimals are similar in that they both allow you to represent parts of the whole — that is, these numbers fall on the number line *between* whole numbers.

In practice, though, sometimes one of these options is more desirable than the other. For example, calculators love decimals but aren't so crazy about fractions. To use your calculator, you may have to change fractions into decimals.

As another example, some units of measurement (such as inches) use fractions, and others (such as meters) use decimals. To change units, you may need to convert between fractions and decimals.

In this section, I show you how to convert back and forth between fractions and decimals.

Changing decimals to fractions

Converting a decimal to a fraction is pretty simple. The only tricky part comes in when you have to reduce the fraction or change it to a mixed number.

Doing a basic decimal-to-fraction conversion

Here's how to convert a decimal to a fraction:

1. **Draw a line (fraction bar) under the decimal and place a 1 underneath it.**

   Suppose you want to turn the decimal 0.3763 into a fraction. Draw a line under 0.3763 and place a 1 underneath it:

   $$\frac{0.3763}{1}$$

   This number looks like a fraction, but technically it isn't one because the top number (the numerator) is a decimal.

2. **Move the decimal point one place to the right and add a 0 after the 1.**

   $$\frac{3.763}{10}$$

3. **Repeat Step 2 until the decimal point moves all the way to the right and you can drop the decimal point entirely.**

   In this case, this is a three-step process:

   $$\frac{37.63}{100} = \frac{376.3}{1{,}000} = \frac{3{,}763}{10{,}000}$$

   As you can see in the last step, the decimal point in the numerator moves all the way to the end of the number, so dropping the decimal point is okay.

   Note: Moving a decimal point one place to the right is the same thing as multiplying a number by 10. When you move the decimal point four places in this problem, you're essentially multiplying the 0.3763 and the 1 by 10,000. Notice that the number of digits after the decimal point in the original decimal is equal to the number of 0s that end up following the 1.

4. **If necessary, change the resulting fraction to a mixed number and/or reduce it.**

   The fraction $\frac{3,763}{10,000}$ may seem like a large number, but it's smaller than 1 because the numerator is less than the denominator (bottom number). So this fraction is a proper fraction and can't be changed into a mixed number (see Chapter 4 to find out more about mixed numbers). This fraction is also in the lowest possible terms, so this problem is complete.

Mixing numbers and reducing fractions

In some cases, you may have to reduce a fraction to lowest terms after you convert it. Reducing in these cases is usually easy, because no matter how large the denominator is, it's a multiple of 10, which has only two prime factors: 2 and 5 (see Chapter 1 for more on prime factors).

When you're converting a decimal to a fraction, if the decimal ends in an even number or in 5, you can reduce the fraction; otherwise, you can't. Also, if you have a number greater than 0 somewhere before the decimal point, you'll end up with a mixed number.

Suppose you want to convert the decimal 12.16 to a fraction. Follow the steps in the proceeding section:

$$\frac{12.16}{1} = \frac{121.6}{10} = \frac{1,216}{100}$$

This time, the numerator is greater than the denominator, so $\frac{1,216}{100}$ is an improper fraction and must be changed to a mixed number.

After you've changed a decimal to an improper fraction, you can convert it to a mixed number by "pulling out" the original whole-number part of the decimal from the fraction.

In this case, the original decimal was 12.16, so its whole number part was 12. Doing the following is perfectly okay:

$$\frac{1,216}{100} = 12\frac{16}{100}$$

This trick works only when you've changed a decimal to a fraction — don't try it with other fractions, or you'll get a wrong answer.

The mixed number $12\frac{16}{100}$ can be also be reduced as follows (if this last step seems unclear, check out Chapter 4 for more on reducing fractions):

$$12\frac{16}{100} = 12\frac{8}{50} = 12\frac{4}{25}$$

Changing fractions to decimals

Converting fractions to decimals isn't difficult, but to do it, you need to know about decimal division. If you need to get up to speed on this, check out "Dividing decimals" earlier in this chapter.

Here's how to convert a fraction to a decimal:

1. **Set up the fraction as a long division problem, dividing the numerator (top number) by the denominator (bottom number).**
2. **Attach enough trailing zeros to the numerator so that you can continue dividing until you find that the answer is either a *terminating decimal* or a *repeating decimal.***

Don't worry. I explain terminating and repeating decimals next.

The last stop: Terminating decimals

Sometimes, when you divide the numerator of a fraction by the denominator, the division eventually works out evenly. The result is a *terminating decimal.*

For example, suppose you want to find out how to represent $\frac{7}{16}$ as a decimal. First, I attach trailing zeros and then proceed to divide normally:

$$
\begin{array}{r}
0.4375 \\
16\overline{)7.0000} \\
\underline{64} \\
60 \\
\underline{-48} \\
120 \\
\underline{-112} \\
80 \\
\underline{-80} \\
0
\end{array}
$$

The division works out evenly, so the answer is a terminating decimal. Therefore, $\frac{7}{16} = 0.4375$.

The endless ride: Repeating decimals

Sometimes when you try to convert a fraction to a decimal, the division doesn't work out evenly. The result is a *repeating decimal* — that is, a decimal that cycles through the same number pattern forever.

For example, to change $\frac{2}{3}$ to a decimal, begin by dividing 2 by 3. As in the preceding section, start out by adding trailing zeros and see where it leads:

$$
\begin{array}{r}
0.666 \\
3\overline{)2.000} \\
\underline{18} \\
20 \\
\underline{-18} \\
20 \\
\underline{-18} \\
2
\end{array}
$$

At this point, you still haven't found an exact answer. But you may notice that a repeating pattern has developed in the division. No matter how many trailing zeros you attach to the number 2, the same pattern will continue forever. This answer, 0.666..., is an example of a repeating decimal. You can write $\frac{2}{3}$ as

$$\frac{2}{3} = 0.\overline{6}$$

The bar over the 6 means that in this decimal, the number 6 repeats forever. You can represent many simple fractions as repeating decimals. In fact, *every* fraction can be represented either as a repeating decimal or as a terminating decimal — that is, as an ordinary decimal that ends.

Now suppose you want to find the decimal representation of $\frac{5}{11}$. Here's how this problem plays out:

$$\begin{array}{r} 0.4545 \\ 11\overline{)5.0000} \\ \underline{44} \\ 60 \\ \underline{-55} \\ 50 \\ \underline{-44} \\ 60 \\ \underline{-55} \\ 5 \end{array}$$

This time, the pattern repeats every other number — 4, then 5, then 4 again, and then 5 again, forever. Attaching more trailing zeros to the original decimal will only string out this pattern indefinitely. So you can write

$$\frac{5}{11} = 0.\overline{45}$$

This time, the bar is over both the 4 and the 5, telling you that these two numbers alternate forever.

Repeating decimals are an oddity, but they aren't hard to work with. In fact, as soon as you can show that the decimal division is repeating, you've found your answer. Just remember to place the bar only over the numbers that keep on repeating.

Chapter 6

Puzzling Out Percents

In This Chapter

- Understanding what percents are
- Converting percents back and forth between decimals and fractions
- Solving both simple and difficult percent problems
- Using the percent circle to solve three types of percent problems

Like whole numbers and decimals, percents are a way to talk about parts of a whole. The word *percent* literally means "for 100," but in practice, it means closer to "out of 100." For example, suppose a school has exactly 100 children — 50 girls and 50 boys. You can say that "50 out of 100" children are girls — or you can shorten it to simply "50 percent." Even shorter than that, you can use the symbol %, which means *percent.*

Saying that 50% of the students are girls is the same as saying that $\frac{1}{2}$ of them are girls. Or if you prefer decimals, it's the same thing as saying that 0.5 of all the students are girls. This example shows you that percents, like fractions and decimals, are just another way of talking about parts of the whole. In this case, the whole is the total number of children in the school. Whether you're talking about cake, a dollar, or a group of children, 50% is still half, 25% is still one-quarter, 75% is still three-quarters, and so on.

In this chapter, I show you how to work with percents. Because percents resemble decimals, I first show you how to convert numbers back and forth between percents and decimals. Next, I show you how to convert back and forth between percents and fractions. When you understand how conversions work, I show you the three basic types of percent problems, plus a method that makes the problems simple.

Understanding Percents Greater than 100%

100% means "100 out of 100" — in other words, everything. What about percentages more than 100%? Well, sometimes percentages like these don't make sense. For example, you can't spend more than 100% of your time playing basketball no matter how much you love the sport; 100% is all the time you have.

But lots of times, percentages larger than 100% are perfectly reasonable. For example, suppose I own a hot dog wagon and I sell the following:

10 hot dogs in the morning

30 hot dogs in the afternoon

The number of hot dogs I sell in the afternoon is 300% of the number I sold in the morning. That's three times as many.

Here's another way of looking at this: I sell 20 more hot dogs in the afternoon than in the morning, so this is a *200% increase* in the afternoon — 20 is twice as many as 10.

Spend a little time thinking about this example until it makes sense. You visit some of these ideas again in Chapter 7 when I show you how to do word problems involving percents.

Converting to and from Percents, Decimals, and Fractions

To solve many percent problems, you need to change the percent to either a decimal or a fraction. Then you can apply what you know about solving decimal and fraction problems. That's why I show you how to convert to and from percents before I show you how to solve percent problems.

Percents and decimals are very similar ways of expressing parts of a whole. This similarity makes converting percents to decimals and vice versa mostly a matter of moving the decimal point.

Percents and fractions both express the same idea — parts of a whole — in different ways. So converting back and forth between percents and fractions isn't quite as simple as just moving the decimal point back and forth. In this section, I cover the ways to convert to and from percents, decimals, and fractions, starting with percents to decimals.

Going from percents to decimals

To convert a percent to a decimal, drop the percent sign (%) and move the decimal point two places to the left. That's all there is to it. Remember that in a whole number, the decimal point comes at the end. For example,

2.5% = 0.025

4% = 0.04

36% = 0.36

111% = 1.11

Changing decimals into percents

To convert a decimal to a percent, move the decimal point two places to the right and add a percent sign (%):

0.07 = 7%

0.21 = 21%

0.375 = 37.5%

Switching from percents to fractions

Converting percents to fractions is fairly straightforward. Remember that the word *percent* means "out of 100." So changing percents to fractions naturally involves the number 100.

To convert a percent to a fraction, use the number in the percent as your numerator (top number) and the number 100 as your denominator (bottom number):

$$39\% = \frac{39}{100}$$

$$86\% = \frac{86}{100}$$

$$217\% = \frac{217}{100}$$

As always with fractions, you may need to reduce to lowest terms or convert an improper fraction to a mixed number (flip to Chapter 4 for more on these topics).

In the three examples, $\frac{39}{100}$ can't be reduced or converted to a mixed number. However, $\frac{86}{100}$ can be reduced because the numerator and denominator are both even numbers:

$$\frac{86}{100} = \frac{43}{50}$$

And $\frac{217}{100}$ can be converted to a mixed number because the numerator (217) is greater than the denominator (100):

$$\frac{217}{100} = 2\frac{17}{100}$$

Once in a while, you may start out with a percentage that's a decimal, such as 99.9%. The rule is still the same, but now you have a decimal in the numerator (top number), which most people don't like to see. To get rid of it, move the decimal point one place to the right in both the numerator and the denominator:

$$99.9\% = \frac{99.9}{100} = \frac{999}{1{,}000}$$

Thus, 99.9% converts to the fraction $\frac{999}{1{,}000}$.

Turning fractions into percents

Converting a fraction to a percent is really a two-step process. Here's how to convert a fraction to a percent:

1. **Convert the fraction to a decimal.**

   For example, suppose you want to convert the fraction $\frac{4}{5}$ to a percent. First convert $\frac{4}{5}$ to a decimal by dividing the numerator by the denominator, as shown in Chapter 5:

   $$\begin{array}{r} 0.8 \\ 5\overline{)4.0} \\ \underline{4\,0} \\ 0 \end{array}$$

2. **Convert this decimal to a percent.**

   Convert 0.8 to a percent by moving the decimal point two places to the right and adding a percent sign (as I show you earlier in "Changing decimals into percents").

   $0.8 = 80\%$

Now suppose you want to convert the fraction $\frac{5}{8}$ to a percent. Follow these steps:

1. **Convert $\frac{5}{8}$ to a decimal by dividing the numerator by the denominator:**

   $$\begin{array}{r} 0.625 \\ 8\overline{)5.000} \\ \underline{4\,8} \\ 20 \\ \underline{-16} \\ 40 \\ \underline{-40} \\ 0 \end{array}$$

   Therefore, $\frac{5}{8} = 0.625$.

2. **Convert 0.625 to a percent by moving the decimal point two places to the right and adding a percent sign (%):**

   $0.625 = 62.5\%$

Solving Percent Problems

When you know the connection between percents and fractions, which I discuss earlier in "Converting to and from

Percents, Decimals, and Fractions," you can solve a lot of percent problems with a few simple tricks. Others, however, require a bit more work. In this section, I show you how to tell an easy percent problem from a tough one, and I give you the tools to solve them all.

Figuring out simple percent problems

A lot of percent problems turn out to be easy when you give them a little thought. In many cases, just remember the connection between percents and fractions, and you're halfway home:

- **Finding 100% of a number:** Remember that 100% means the whole thing, so 100% of any number is simply the number itself:

  100% of 91 is 91

  100% of 732 is 732

- **Finding 50% of a number:** Remember that 50% means half, so to find 50% of a number, just divide it by 2:

  50% of 88 is 44

  50% of 7 is $\frac{7}{2}$ (or $3\frac{1}{2}$ or 3.5)

- **Finding 25% of a number:** Remember that 25% equals $\frac{1}{4}$, so to find 25% of a number, divide it by 4:

  25% of 88 is 22

  25% of 15 is $\frac{15}{4}$ (or $3\frac{3}{4}$ or 3.75)

- **Finding 20% of a number:** Finding 20% of a number is handy if you like the service you've received in a restaurant, because a good tip is 20% of the check. Because 20% equals $\frac{1}{5}$, you can find 20% of a number by dividing it by 5. But here's an easier way: To find 20% of a number, move the decimal point one place to the left and double the result:

  20% of 80 = $8 \cdot 2 = 16$

  20% of 300 = $30 \cdot 2 = 60$

  20% of 41 = $4.1 \cdot 2 = 8.2$

- **Finding 200%, 300%, and so on of a number:** Working with percents that are multiples of 100 is easy. Just drop the two 0s and multiply by the number that's left:

  200% of $7 = 2 \cdot 7 = 14$

  300% of $10 = 3 \cdot 10 = 30$

  $1{,}000\%$ of $45 = 10 \cdot 45 = 450$

  (See the earlier "Understanding Percents Greater than 100%" section for details on what having more than 100% really means.)

Deciphering more-difficult percent problems

You can solve a lot of percent problems using the tricks I show you earlier in this chapter. But what about this problem?

35% of $80 = ?$

Ouch — this time, the numbers you're working with aren't so friendly. When the numbers in a percent problem become a little more difficult, the tricks no longer work, so you want to know how to solve *all* percent problems.

Here's how to find any percent of any number:

1. **Change the word *of* to a multiplication sign and the percent to a decimal (as I show you earlier in this chapter).**

   Changing the word *of* to a multiplication sign is a simple example of turning words into numbers, as I discuss in Chapters 3 and 7. This change turns something unfamiliar into a form that you know how to work with.

   Suppose you want to find 35% of 80. Here's how you start:

   35% of $80 = 0.35 \cdot 80$

2. **Solve the problem using decimal multiplication (see Chapter 5 for details).**

Here's what the example looks like:

$$\begin{array}{r} 0.35 \\ \times\ 80 \\ \hline 28.00 \end{array}$$

So 35% of 80 is 28.

As another example, suppose you want to find 12% of 31. Again, start by changing the percent to a decimal and the word *of* to a multiplication sign:

$$12\% \text{ of } 31 = 0.12 \cdot 31$$

Now you can solve the problem with decimal multiplication:

$$\begin{array}{r} 0.12 \\ \times\ 31 \\ \hline 12 \\ +\ 360 \\ \hline 3.72 \end{array}$$

So 12% of 31 is 3.72.

Applying Percent Problems

In the preceding section, "Solving Percent Problems," I give you a few ways to find any percent of any number. This type of percent problem is the most common — that's why it gets top billing.

But percents are commonly used in a wide range of business applications such as banking, real estate, payroll, and taxes. (I show you some real-world applications when I discuss word problems in Chapter 7.) And depending on the situation, two other common types of percent problems may present themselves.

In this section, I show you these two additional types of percent problems and how they relate to the type you now know how to solve. I also give you the percent circle, a simple tool to make quick work of all three types.

Identifying the three types of percent problems

Earlier in this chapter, I show you how to solve problems that look like this:

50% of 2 is ?

And the answer, of course, is 1. (See "Solving Percent Problems" for details on how to get this answer.) Given two pieces of information — the percent and the number to start with — you can figure out what number you end up with.

Now suppose instead that I leave out the percent but give you the starting and ending numbers:

?% of 2 is 1

You can still fill in the blank without too much trouble. Similarly, suppose that I leave out the starting number but give the percent and the ending number:

50% of ? is 1

Again, you can fill in the blank.

If you get this basic idea, you're ready to solve percent problems. When you boil them down, nearly all percent problems are like one of the three types I show in Table 6-1.

Table 6-1 The Three Main Types of Percent Problems

Problem Type	*What to Find*	*Example*
Type 1	The ending number	50% of 2 is *what?*
Type 2	The percentage	*What* percent of 2 is 1?
Type 3	The starting number	50% of *what* is 1?

In each case, the problem gives you two of the three pieces of information, and your job is to figure out the remaining piece. In the next section, I give you a simple tool to help you solve all three of these types of percent problems.

Introducing the percent circle

The *percent circle* is a simple visual aid that helps you make sense of percent problems so that you can solve them easily. The trick to using a percent circle is to write information in it correctly. For example, Figure 6-1 shows how to record the information that 50% of 2 is 1.

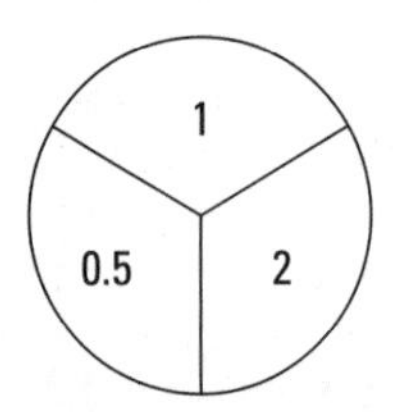

Figure 6-1: In the percent circle, the ending number is at the top, the percentage is at the left, and the starting number is at the right.

Notice that as I fill in the percent circle, I change the percentage, 50%, to its decimal equivalent, 0.5 (for more on changing percents to decimals, see "Going from percents to decimals" earlier in this chapter).

Here are the two main features of the percent circle:

- When you multiply the two bottom numbers together, they equal the top number:

  $$0.5 \cdot 2 = 1$$

- If you make a fraction out of the top number and either bottom number, that fraction equals the other bottom number:

  $$\frac{1}{2} = 0.5 \text{ and } \frac{1}{0.5} = 2$$

These features are the heart and soul of the percent circle. They enable you to solve any of the three types of percent problems quickly and easily.

Most percent problems give you enough information to fill in two of the three sections of the percent circle. But no matter which two sections you fill in, you can find out the number in the third section.

Finding the ending number

Suppose you want to find out the answer to this problem:

What is 75% of 20?

You're given the percent and the starting number and asked to find the ending number. To use the percent circle on this problem, fill in the information as Figure 6-2 shows.

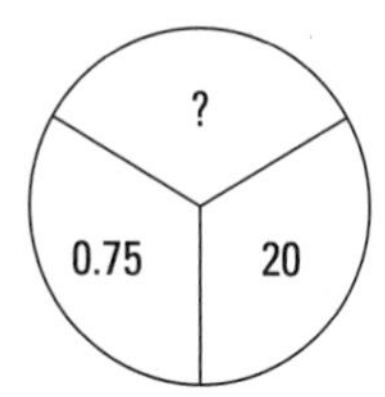

Figure 6-2: Putting 75% of 20 into a percent circle.

Because 0.75 and 20 are both bottom numbers in the circle, multiply them to get the answer:

$$\begin{array}{r} 0.75 \\ \underline{\times\ 20} \\ 15.00 \end{array}$$

So 75% of 20 is 15.

As you can see, this method is essentially the same one I show you earlier in this chapter in "Deciphering more-difficult percent problems," where you translate the word *of* as a multiplication sign. You still use multiplication to get your answer, but with the percent circle, you're less likely to get confused.

Discovering the percentage

In the second type of problem, I give you both the starting and ending numbers, and I ask you to find the percentage. Here's an example:

What percent of 50 is 35?

In this case, the starting number is 50 and the ending number is 35. Set up the problem on the percent circle as Figure 6-3 shows.

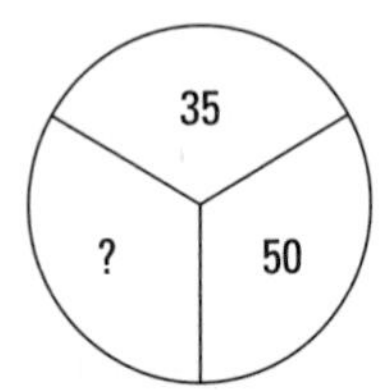

Figure 6-3: Determining what percent of 50 is 35.

This time, 35 is above 50, so make a fraction out of these two numbers:

$$\frac{35}{50}$$

This fraction *is your answer,* and all you have to do is convert the fraction to a percent as I discuss earlier in this chapter in "Turning fractions into percents." First, convert $\frac{35}{50}$ to a decimal:

$$\begin{array}{r} 0.7 \\ 50\overline{)35.0} \\ \underline{350} \\ 0 \end{array}$$

Now convert 0.7 to a percent:

$$0.7 = 70\%$$

Tracking down the starting number

In the third type of problem, you get the percentage and the ending number, and you have to find the starting number. For example,

15% of what number is 18?

This time, the percentage is 15% and the ending number is 18, so fill in the percent circle as Figure 6-4 shows.

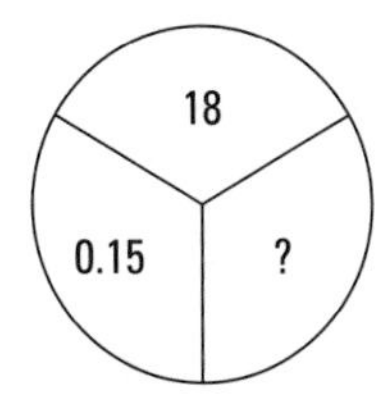

Figure 6-4: Working out the answer to "15% of what number is 18?"

Because 18 is above 0.15 in the circle, make a fraction out of these two numbers:

$$\frac{18}{0.15}$$

This fraction *is your answer;* you just need to convert to a decimal as I show you in Chapter 5. Divide 18 by 0.15:

$$\begin{array}{r} 120. \\ 15.\overline{)1800.} \\ \underline{15} \\ 30 \\ \underline{-30} \\ 0 \end{array}$$

In this case, the "decimal" you find is the whole number 120, so 15% of 120 is 18.

Chapter 7

Fraction, Decimal, and Percent Word Problems

In This Chapter

- Adding and subtracting fractions, decimals, and percents in word equations
- Translating the word *of* as multiplication
- Changing percents to decimals in word problems
- Tackling business problems involving percent increase and decrease

In Chapter 3, I show you how to solve word problems by setting up word equations that use the Big Four operations. In this chapter, you see how to extend these skills to solve word problems with fractions, decimals, and percents. You discover how to solve percent problems by setting up word equations and changing the percent to a decimal. Finally, I show you how to handle practical problems of percent increase and decrease, such as raises and salaries, costs and discounts, and amounts before and after taxes.

Adding and Subtracting Parts of the Whole

Certain word problems involving fractions, decimals, and percents are really just problems in adding and subtracting. For instance, you may add fractions, decimals, or percents in a variety of real-world settings that rely on weights and measures, such as cooking and carpentry.

Sharing a pizza: Fractions

You may have to add or subtract fractions in problems that involve splitting up part of a whole. For example, consider the following:

> Joan ate $\frac{1}{6}$ of a pizza, Tony ate $\frac{1}{4}$, and Sylvia ate $\frac{1}{3}$. What fraction of the pizza was left when they were finished?

Just jot down the information that's given in this problem as word equations:

$$\text{Joan} = \frac{1}{6} \quad \text{Tony} = \frac{1}{4} \quad \text{Sylvia} = \frac{1}{3}$$

These fractions are part of one total pizza. To solve the problem, you need to find out how much all three people ate, so form the following word equation:

$$\text{All three} = \text{Joan} + \text{Tony} + \text{Sylvia}$$

Now you can substitute as follows:

$$\text{All three} = \frac{1}{6} + \frac{1}{4} + \frac{1}{3}$$

Here's how you can add these fractions (see Chapter 4 for additional ways):

$$\text{All three} = \frac{2}{12} + \frac{3}{12} + \frac{4}{12} = \frac{9}{12} = \frac{3}{4}$$

However, the question asks what fraction of the pizza was left after they finished, so you have to subtract that amount from the whole:

$$1 - \frac{3}{4} = \frac{1}{4}$$

Thus, the three people left $\frac{1}{4}$ of a pizza.

Buying by the pound: Decimals

You frequently work with decimals when dealing with money, metric measurements, and food sold by the pound. The following problem requires you to add and subtract decimals, which I discuss in Chapter 5. Even though the decimals may look intimidating, this problem is fairly simple to set up:

> Antonia bought 4.53 pounds of beef and 3.1 pounds of lamb. Lance bought 5.24 pounds of chicken and 0.7 pounds of pork. Which of them bought more meat, and how much more?

To solve this problem, you first find out how much each person bought:

Antonia = 4.53 + 3.1 = 7.63

Lance = 5.24 + 0.7 = 5.94

You can already see that Antonia bought more than Lance. To find how much more, subtract:

$7.63 - 5.94 = 1.69$

So Antonia bought 1.69 pounds more than Lance.

Splitting the vote: Percents

When percents represent answers in polls, votes in an election, or portions of a budget, the total usually has to add up to 100%. In real life, you may see such info organized as a pie chart (which I discuss in Chapter 12). Solving problems about this kind of information often involves nothing more than adding and subtracting percents. Here's an example:

> In a recent mayoral election, five candidates were on the ballot. Faber won 39% of the vote, Gustafson won 31%, Ivanovich won 18%, Dixon won 7%, Obermayer won 3%, and the remaining votes went to write-in candidates. What percentage of voters wrote in their selections?

The candidates were in a single election, so all the votes have to total 100%. The first step here is just to add the five percentages:

$39\% + 31\% + 18\% + 7\% + 3\% = 98\%$

Then subtract that value from 100%:

$100\% - 98\% = 2\%$

Because 98% of voters voted for one of the five candidates, the remaining 2% wrote in their selections.

Multiplying Fractions in Everyday Situations

In word problems, the word *of* almost always means multiplication. So whenever you see the word *of* following a fraction, decimal, or percent, you can usually replace it with a times sign.

When you think about it, *of* means multiplication even when you're not talking about fractions. For example, when you point to an item in a store and say, "I'll take three of those," in a sense you're saying, "I'll take that one multiplied by three."

Buying less than advertised

After you understand that the word *of* means multiplication, you have a powerful tool for solving word problems. For instance, you can figure out how much you'll spend if you don't buy food in the quantities listed on the signs. Here's an example:

> If beef costs \$4 a pound, how much does $\frac{5}{8}$ of a pound cost?

Here's what you get if you simply change the *of* to a multiplication sign:

$$\frac{5}{8} \times 1 \text{ pound of beef}$$

That's how much beef you're buying. However, you want to know the cost. Because the problem tells you that 1 pound = \$4, you can replace *1 pound of beef* with *\$4:*

$$= \frac{5}{8} \cdot \$4$$

Now you have an expression you can evaluate. Use the rules of multiplying fractions from Chapter 4 and solve:

$$= \frac{5 \cdot \$4}{8 \cdot 1} = \frac{\$20}{8}$$

This fraction reduces to $\frac{\$5}{2}$. However, the answer looks weird because dollars are usually expressed in decimals, not fractions. So convert this fraction to a decimal:

$$\frac{\$5}{2} = \$5 \div 2 = \$2.5$$

At this point, recognize that \$2.5 is more commonly written as \$2.50, and you have your answer.

Computing leftovers

Sometimes, when you're sharing something such as a pie, not everyone gets to it at the same time. The eager pie-lovers snatch the first slice, not bothering to divide the pie into equal servings, and the people who were more patient cut their own portions from what's left over. When someone takes a part of the leftovers, you can do a bit of multiplication to see how much of the *whole* pie that portion represents.

Consider the following example:

> Jerry bought a pie and ate $\frac{1}{5}$ of it. Then his wife Doreen ate $\frac{1}{6}$ of what was left. How much of the total pie was left?

To solve this problem, begin by jotting down what the first sentence tells you:

$$\text{Jerry} = \frac{1}{5}$$

Doreen ate part of what was left, so write a word equation that tells you how much of the pie was left after Jerry ate. He started with a whole pie, so subtract his portion from 1:

$$\text{Pie left after Jerry} = 1 - \frac{1}{5} = \frac{4}{5}$$

Doreen ate $\frac{1}{6}$ of this amount. This next answer tells you how much of the *whole* pie Doreen ate:

$$\text{Doreen} = \frac{1}{6} \cdot \frac{4}{5} = \frac{4}{30}$$

To make the numbers a little smaller before you go on, notice that you can reduce the fraction:

$$\text{Doreen} = \frac{2}{15}$$

Now you know how much Jerry and Doreen both ate, so you can add these amounts together:

$$\text{Jerry} + \text{Doreen} = \frac{1}{5} + \frac{2}{15}$$

Add the fractions (as I show you in Chapter 4):

$$= \frac{3}{15} + \frac{2}{15} = \frac{5}{15}$$

This fraction reduces to $\frac{1}{3}$. Now you know that Jerry and Doreen ate $\frac{1}{3}$ of the pie, but the problem asks you how much is left. So finish up with some subtraction and write the answer:

$$1 - \frac{1}{3} = \frac{2}{3}$$

The amount of pie left over was $\frac{2}{3}$.

Multiplying Decimals and Percents in Word Problems

The word *of* usually means multiplication in word problems involving decimals and percents, not just fractions.

You can easily solve word problems involving percents by changing the percents into decimals.

Figuring out how much money is left

One common type of problem gives you a starting amount — and a bunch of other information — and then asks you to figure out how much you end up with. Here's an example:

Maria's grandparents gave her $125 for her birthday. She put 40% of the money in the bank, spent 35% of what was left on a pair of shoes, and then spent the rest on a dress. How much did the dress cost?

Start at the beginning, forming a word equation to find out how much money Maria put in the bank:

Money in bank = 40% of $125

To solve this word equation, change the percent to a decimal and the word *of* to a multiplication sign; then multiply:

Money in bank = 0.4 · $125 = $50

Pay special attention to whether you're calculating how much of something was used up or how much of something is left over. If you need to work with the portion that remains, you often have to subtract the amount used from the amount you started with.

Because Maria started with $125, she had $75 left to spend:

Money left to spend = Money from grandparents – Money in bank

= $125 – $50 = $75

The problem then says that she spent 35% of *this* amount on a pair of shoes. Again, change the percent to a decimal and the word *of* to a multiplication sign:

Shoes = 35% of $75 = 0.35 · $75 = $26.25

She spent the rest of the money on a dress, so

Dress = $75 – $26.25 = $48.75

Therefore, Maria spent $48.75 on the dress.

Finding out how much you started with

Some problems give you the amount that you end up with and ask you to find out how much you started with. In general, these problems are harder because you're not used to thinking backwards. Here's an example, and it's kind of a tough one, so fasten your seat belt:

> Maria received some birthday money from her aunt. She put her usual 40% in the bank, spent 75% of the rest on a purse, and when she was done, she had $12 left to spend on dinner. How much did her aunt give her?

Notice that the only dollar amount in the problem comes after the two percent amounts. The problem tells you that she *ends up with* $12 after two transactions — putting money in the bank and buying a purse — and asks you to find out how much you started with.

To solve this problem, set up two word equations to describe the two transactions:

Money from aunt – money for bank = money after bank

Money after bank – money for purse = $12

Notice what these two word equations are saying. The first tells you that Maria took the money from her aunt, subtracted some money to put in the bank, and left the bank with a new amount of money, which I'm calling *money after bank*. The second word equation starts where the first leaves off. It tells you that Maria took the money left over from the bank, subtracted some money for a purse, and ended up with $12.

This second equation already has an amount of money filled in, so start here. To solve this problem, realize that Maria spent 75% of her money *at that time* on the purse — that is, 75% of the money she still had after the bank:

100% of money after bank – 75% of money after bank = $12

Because 100% – 75% = 25%, here's an even better way to write this equation:

25% · money after bank = $12

Before moving on, make sure you understand the steps that have brought you here. The percent circle (see Chapter 6) tells you that $12 · 0.25 gives you the remaining amount of $48.

Okay, you know how much money Maria had after the bank, so you can plug this number into the first equation:

Money from aunt – money for bank = $48

Now you can use the same type of thinking to solve this equation. First, Maria placed 40% of the money from her aunt in the bank:

100% of money from aunt – 40% of money from aunt = $48

Now, because 100% – 40% = 60%, rewrite the word equation:

60% · money from aunt = $48

At this point, you can use the percent circle (see Chapter 6) to solve the equation. In this case, the percent circle tells you that $48 ÷ 0.6 gives you the remaining amount of $80. So Maria's aunt gave her $80 for her birthday.

Handling Percent Increases and Decreases in Word Problems

Word problems that involve increasing or decreasing by a percentage add a final spin to percent problems. A percent-increase problem may involve calculating the amount of a salary plus a raise, the cost of merchandise plus tax, or an amount of money plus interest or dividend. A percent decrease problem may involve the amount of a salary minus taxes or the cost of merchandise minus a discount.

These problems are similar to the ones in "Multiplying Decimals and Percents in Word Problems." But people often get thrown by the language of these problems — which, by the way, is the language of business — so I want to give you some practice solving them.

Raking in the dough: Finding salary increases

A little common sense tells you that the words *salary increase* or *raise* means more money, so get ready to do some addition. Here's an example problem:

> Alison's salary was $40,000 last year, and at the end of the year, she received a 5% raise. What will she earn this year?

To solve this problem, first realize that Alison got a raise. So whatever she makes this year, it'll be more than she made last year. The key to setting up this type of problem is to think of percent increase as "100% of last year's salary plus 5% of last year's salary." Here's the word equation:

This year's salary = 100% of last year's salary + 5% of last year's salary

Now you can just add the percentages:

This year's salary = (100% + 5%) of last year's salary

= 105% of last year's salary

Change the percent to a decimal and the word *of* to a multiplication sign; then fill in the amount of last year's salary:

This year's salary = 1.05 · $40,000

Now you're ready to multiply:

This year's salary = $42,000

So Alison's new salary is $42,000.

Earning interest on top of interest

The word *interest* means more money. When you receive interest from the bank, you get more money. And when you pay interest on a loan, you pay more money. Sometimes people earn interest on the interest they earned earlier, which makes the dollar amounts grow even faster. Here's an example problem:

> Bethany placed $9,500 in a one-year CD that paid 4% interest. The next year, she rolled this over into a bond that paid 6% per year. How much did Bethany earn on her investment in those two years?

This problem involves interest, so it's a problem in percent increase, only this time, you have to deal with two transactions. Take them one at a time.

The first transaction is a percent increase of 4% on $9,500. The following word equation makes sense:

$$\text{Money after first year} = 100\%\text{ of initial deposit} + 4\%\text{ of initial deposit}$$
$$= (100\% + 4\%)\text{ of }\$9{,}500$$
$$= 104\%\text{ of }\$9{,}500$$

Now change the percent to a decimal and the word *of* to a multiplication sign:

$$\text{Money after first year} = 1.04 \cdot \$9{,}500$$
$$= \$9{,}880$$

At this point, you're ready for the second transaction. This is a percent increase of 6% on $9,880:

$$\text{Final amount} = 106\%\text{ of }\$9{,}880$$

Again, change the percent to a decimal and the word *of* to a multiplication sign:

$$\text{Final amount} = 1.06 \cdot \$9{,}880 = \$10{,}472.80$$

Then subtract the initial deposit from the final amount to find her earnings:

$$\text{Earnings} = \text{final amount} - \text{initial deposit}$$
$$= \$10{,}472.80 - \$9{,}500 = \$972.80$$

So Bethany earned $972.80 on her investment.

Getting a deal: Calculating discounts

When you hear the word *discount* or *sale price,* think of subtraction. Here's an example problem:

> Greg has his eye on a television with a listed price of \$2,100. The salesman offers him a 30% discount if he buys it today. What will the television cost with the discount?

In this problem, you need to realize that the discount lowers the price of the television, so you have to subtract:

Sale price = 100% of regular price – 30% of regular price

Now subtract percentages:

Sale price = (100% – 30%) of regular price

= 70% of regular price

At this point, you can fill in the details:

Sale price = 0.7 · \$2,100 = \$1,470

Thus, the television will cost \$1,470 with the discount.

Chapter 8

Using Variables in Algebraic Expressions

In This Chapter

- Understanding how a variable such as x stands for a number
- Using substitution to evaluate an algebraic expression
- Identifying and rearranging the terms in any algebraic expression
- Simplifying algebraic expressions

People use algebra to solve problems that are just too difficult for ordinary arithmetic. And because number-crunching is so much a part of the modern world, algebra is everywhere (even if you don't see it): architecture, engineering, medicine, statistics, computers, business, chemistry, physics, biology, and of course, higher math. Anywhere that numbers are useful, algebra is there. That's why virtually every college insists that you leave (or enter) with at least a passing familiarity with algebra.

In this chapter, I introduce (or reintroduce) you to algebra in a way that's bound to make it seem a little friendlier. Then I show you how algebraic expressions are similar to and different from the arithmetic expressions that you're used to working with. (For a refresher on arithmetic expressions, see Chapter 2.)

Variables: X Marks the Spot

In math, x stands for a number — any number. Any letter that you use to stand for a number is a *variable,* which means that its value can vary — that is, its value is uncertain. In contrast, a number in algebra is often called a *constant,* because its value is fixed.

Sometimes, you have enough information to find out the identity of x. For example, consider the following:

$$2 + 2 = x$$

Obviously, in this equation, x stands for the number 4. But other times, the number that x stands for stays shrouded in mystery. For example,

$$x > 5$$

In this inequality, x stands for some number greater than 5 — maybe 6, maybe $7\frac{1}{2}$, maybe 542.002.

Expressing Yourself with Algebraic Expressions

In Chapter 2, I introduce you to *arithmetic expressions:* strings of numbers and operators that you can evaluate or place on one side of an equation. For example, here are some arithmetic expressions:

$$2 + 3 \qquad 7 \cdot 1.5 - 2 \qquad 2^4 - |-4| - \sqrt{100}$$

In this chapter, I introduce you to another type of mathematical expression: the algebraic expression. An *algebraic expression* is any string of mathematical symbols that you can place on one side of an equation and that includes at least one variable. Here are a few examples of algebraic expressions:

$$5x \qquad -5x + 2 \qquad x^2y - y^2x + \frac{z}{3} - xyz + 1$$

As you can see, the difference between arithmetic and algebraic expressions is simply that an algebraic expression includes at least one variable.

An algebraic expression can have any number of variables, but you usually don't work with expressions that have more than three. You can use any letter as a variable, but *x, y,* and *z* tend to get a lot of mileage.

In this section, I demonstrate how to evaluate an algebraic expression by substituting the values of its variables. Then I show you how to separate an algebraic expression into one or more terms and how to identify the coefficient and the variable part of each term.

Evaluating algebraic expressions

Evaluating an expression means finding the value of that expression as a single number. To evaluate an algebraic expression, you need to know the numerical value of every variable. For each variable in the expression, substitute the number that it stands for; then evaluate the expression.

Knowing how to evaluate arithmetic expressions (see Chapter 2) comes in handy for evaluating algebraic expressions. For example, suppose you want to evaluate the following expression:

$$4x - 7$$

Note that this expression contains the variable x, which is unknown, so the value of the whole expression is also unknown. But suppose you know that $x = 2$. To evaluate the expression, substitute 2 for x everywhere it appears in the expression:

$$4(2) - 7$$

After you make the substitution, you're left with an arithmetic expression, so you can finish your calculations to evaluate the expression:

$$= 8 - 7 = 1$$

So given $x = 2$, the algebraic expression $4x - 7 = 1$.

Now suppose you want to evaluate the following expression, where $x = 4$:

$$2x^2 - 5x - 15$$

The first step is to substitute 4 for x everywhere this variable appears in the expression:

$$2(4^2) - 5(4) - 15$$

Now evaluate according to the order of operations (which I explain in Chapter 2). You do powers first, so begin by evaluating the exponent 4^2, which equals $4 \cdot 4$, or 16:

$$= 2(16) - 5(4) - 15$$

Now proceed to the multiplication, moving from left to right:

$$= 32 - 5(4) - 15$$
$$= 32 - 20 - 15$$

Then evaluate the subtraction, again from left to right:

$$= 12 - 15$$
$$= -3$$

So given $x = 4$, the algebraic expression $2x^2 - 5x - 15 = -3$.

You aren't limited to expressions of only one variable when using substitution. As long as you know the value of every variable in the expression, you can evaluate algebraic expressions with any number of variables. For example, suppose you want to evaluate this expression:

$$3x^2 + 2xy - xyz$$

To evaluate it, you need values of all three variables. Say that $x = 3$, $y = -2$, and $z = 5$. The first step is to substitute the equivalent value for each of the three variables wherever you find them:

$$3(3^2) + 2(3)(-2) - 3(-2)(5)$$

Now use the rules for order of operations (from Chapter 2).

$$= 3(9) + 2(3)(-2) - 3(-2)(5)$$
$$= 27 + 2(3)(-2) - 3(-2)(5)$$
$$= 27 + 6(-2) - 3(-2)(5)$$
$$= 27 + -12 - 3(-2)(5)$$
$$= 27 + -12 - (-6)(5)$$
$$= 27 + -12 - (-30)$$
$$= 15 - (-30)$$
$$= 15 + 30$$
$$= 45$$

So given the three values for x, y, and z, the algebraic expression $3x^2 + 2xy - xyz = 45$.

Coming to algebraic terms

A *term* in an algebraic expression is any chunk of symbols set off from the rest of the expression by either addition or subtraction. As algebraic expressions get more complex, they begin to string themselves out in more and more terms. Here are some examples:

Expression	***Number of Terms***	***Terms***
$5x$	One	$5x$
$-5x + 2$	Two	$-5x$ and 2
$x^2y + \frac{z}{3} - xyz + 8$	Four	x^2y, $\frac{z}{3}$, $-xyz$, 8

No matter how complicated an algebraic expression gets, you can always separate it out into one or more terms.

When separating an algebraic expression into terms, group the plus or minus sign with the term that comes right after it.

When a term has a variable, it's called an *algebraic term*. When it doesn't have a variable, it's called a *constant*. For example, look at the following expression:

$$x^2y + \frac{z}{3} - xyz + 8$$

The first three terms are algebraic terms, and the last term is a constant. In algebra, *constant* is just a fancy word for *number.*

Terms are really useful because you can follow rules to move them, combine them, and perform the Big Four operations on them. All these skills are important for solving equations, which I explain in the next chapter.

Making the commute: Rearranging your terms

After you understand how to separate an algebraic expression into terms, you can rearrange the terms in any order you like. Each term moves as a unit.

For example, suppose you begin with the expression $-5x + 2$. You can rearrange the two terms of this expression without changing its value. Notice that each term's sign stays with that term, though dropping the plus sign at the beginning of an expression is customary:

$$-5x + 2 = 2 - 5x$$

Rearranging terms in this way doesn't affect the value of the expression because addition is *commutative* — that is, you can rearrange things that you're adding without changing the answer. (See Chapter 1 for info on the commutative property of addition.)

For example, suppose $x = 3$. Then the original expression and its rearrangement evaluate as follows:

$$\begin{array}{ll} -5x+2 & 2-5x \\ =-5(3)+2 & =2-5(3) \\ =-15+2 & =2-15 \\ =-13 & =-13 \end{array}$$

Rearranging expressions in this way becomes handy later in this chapter, when you simplify algebraic expressions.

As long as each term's sign stays with that term, rearranging the terms in an expression has no effect on its value.

Suppose that $x = 2$ and $y = 3$. Here's how to evaluate the original expression $4x - y + 6$ and two rearrangements:

$4x - y + 6$	$6 + 4x - y$	$-y + 4x + 6$
$= 4(2) - 3 + 6$	$= 6 + 4(2) - 3$	$= -3 + 4(2) + 6$
$= 8 - 3 + 6$	$= 6 + 8 - 3$	$= -3 + 8 + 6$
$= 5 + 6$	$= 14 - 3$	$= 5 + 6$
$= 11$	$= 11$	$= 11$

Identifying the coefficient and variable

Every term in an algebraic expression has a coefficient. The *coefficient* is the signed numerical part of a term in an algebraic expression — that is, the number and the sign (+ or –) that goes with that term. For example, suppose you're working with the following algebraic expression:

$$-4x^3 + x^2 - x - 7$$

The table below shows the four terms of this expression, with each term's coefficient:

Term	***Coefficient***	***Variable***
$-4x^3$	–4	x^3
x^2	1	x^2
$-x$	–1	x
–7	None	None

Notice that the sign associated with the term is part of the coefficient, so the coefficient of $-4x^3$ is –4.

When a term appears to have no coefficient, the coefficient is actually 1. Therefore, the coefficient of x^2 is 1, and the coefficient of $-x$ is –1. When a term appears without a variable, that number with its associated sign is simply a constant.

By the way, when the coefficient of any algebraic term is 0, the expression equals 0 no matter what the variable part looks like:

$0x = 0$ $\quad$ $0xyz = 0$ $\quad$ $0x^3y^4z^{10} = 0$

Identifying similar terms

Similar terms (or *like terms*) are any two algebraic terms that have the same variable part — both the letters and their exponents have to be exact matches. Here are some examples:

Variable Part	*Examples of Similar Terms*		
x	$4x$	$12x$	$99.9x$
x^2	$6x^2$	$-20x^2$	$\frac{8}{3}x^2$
y	y	$1{,}000y$	πy
xy	$-7xy$	$800xy$	$\frac{22}{7}xy$
x^3y^3	$3x^3y^3$	$-111x^3y^3$	$3.14x^3y^3$

In each example, the variable part in all three similar terms is the same. Only the coefficient changes, and the coefficient can be any real number: positive or negative, whole number, fraction, decimal, or even an irrational number such as π.

Considering algebraic terms and the Big Four operations

In this section, I get you up to speed on how to apply the Big Four to algebraic expressions. For now, just think of the ways to work with algebraic expressions as a set of tools that you're collecting for use when you get on the job. You find how useful these tools are in Chapter 9, when you begin solving algebraic equations.

Adding terms

Add similar terms by adding their coefficients and keeping the same variable part.

For example, suppose you have the expression $2x + 3x$. Remember that $2x$ is just shorthand for $x + x$, and $3x$ means simply $x + x + x$. So when you add them up, you get the following:

$$= x + x + x + x + x = 5x$$

As you can see, when the variable parts of two terms are the same, you add these terms by adding their coefficients: $2x + 3x = (2 + 3)x$. The idea here is roughly similar to the idea that 2 apples + 3 apples = 5 apples.

You cannot add dissimilar terms. Here are some cases in which the variables or their exponents are different:

$$2x + 3y \qquad 2yz + 3y \qquad 2x^2 + 3x$$

In these cases, you can't perform the addition. You're faced with a situation that's similar to 2 apples + 3 oranges. Because apples and oranges are different, you can't solve the problem.

Subtracting terms

Subtract similar terms in algebra by finding the difference between their coefficients and keeping the same variable part.

For example, suppose you have $3x - x$. Recall that $3x$ is simply shorthand for $x + x + x$. So doing this subtraction gives you the following:

$$x + x + x - x = 2x$$

No big surprises here. You simply find $(3 - 1)x$. This time, the idea roughly parallels the idea that \$3 – \$1 = \$2.

Here's another example:

$$2x - 5x$$

Again, no problem, as long as you know how to work with negative numbers. Just find the difference between the coefficients:

$$= (2 - 5)x = -3x$$

Similarly, recall that \$2 – \$5 = –\$3 (that is, a debt of 3 dollars).

You cannot subtract dissimilar terms. For example, you can't subtract either of the following:

$$7x - 4y \qquad 7x^2y - 4xy^2$$

As with addition, you can't do subtraction with different variables. Think of this as trying to figure out 7 dollars – 4 pesos. Because the units in this case (dollars versus pesos) are different, you're stuck.

Multiplying terms

You can multiply dissimilar terms. Multiply any two terms by multiplying their coefficients and combining — that is, by collecting or gathering up — all the variables in each term into a single term.

For example, suppose you want to multiply $5x(3y)$. To get the coefficient, multiply $5 \cdot 3$. To get the algebraic part, combine the variables x and y:

$$= 5(3)xy = 15xy$$

Now suppose you want to multiply $2x(7x)$. Again, multiply the coefficients and collect the variables into a single term:

$$= 7(2)xx = 14xx$$

Because x^2 is shorthand for xx, you can write the answer more efficiently:

$$= 14x^2$$

Here's another example. Multiply all three coefficients together and gather up the variables:

$$\begin{aligned} &2x^2(3y)(4xy) \\ &= 2(3)(4)x^2xyy \\ &= 24x^3y^2 \end{aligned}$$

As you can see, the exponent ^{3} that's associated with x is just the count of how many x's appear in the problem. The same is true of the exponent ^{2} associated with y.

A fast way to multiply variables with exponents is to add the exponents together. For example,

$$(x^4y^3)(x^2y^5)(x^6y) = x^{12}y^9$$

In this example, I added the exponents of the *x*'s ($4 + 2 + 6 = 12$) to get the exponent of *x* in the solution. I added the exponents of the *y*'s ($3 + 5 + 1 = 9$) to get the exponent of *y* in the solution.

Dividing terms

It's customary to represent division of algebraic expressions as a fraction instead of using the division sign (÷). So division of algebraic terms really looks like reducing a fraction to lowest terms (see Chapter 4 for more on reducing).

To divide one algebraic term by another, follow these steps:

1. **Make a fraction of the two terms.**

   Suppose you want to divide $3xy$ by $12x^2$. Begin by turning the problem into a fraction:

   $$\frac{3xy}{12x^2}$$

2. **Cancel out factors in the coefficients that are in both the numerator and denominator.**

   In this case, 3 is a factor of both 3 and 12, so you can cancel out a 3. Notice that when the coefficient of *xy* becomes 1, you can drop it:

   $$= \frac{xy}{4x^2}$$

3. **Cancel out any variable that's in both the numerator and denominator.**

   You can break x^2 out as *xx*:

   $$= \frac{xy}{4xx}$$

   Now you can clearly cancel an *x* in both the numerator and denominator:

   $$= \frac{y}{4x}$$

   As you can see, the resulting fraction is really a reduced form of the original.

As another example, suppose you want to divide $-6x^2yz^3$ by $-8x^2y^2z$. Begin by writing the division as a fraction:

$$\frac{-6x^2yz^3}{-8x^2y^2z}$$

First, reduce the coefficients. Notice that because both coefficients were originally negative, you can cancel out both minus signs as well:

$$=\frac{3x^2yz^3}{4x^2y^2z}$$

Now you can begin canceling variables. I do this in two steps as before, first writing out the x's, y's, and z's:

$$=\frac{3xxyzzz}{4xxyyz}$$

At this point, just cross out any occurrence of a variable that appears in both the numerator and denominator:

$$=\frac{3zz}{4y}$$

$$=\frac{3z^2}{4y}$$

You can't cancel out variables or coefficients if either the numerator or denominator has more than one term in it. Terms are chunks separated by plus or minus signs (see the earlier section "Coming to algebraic terms" for details).

Simplifying Algebraic Expressions

Simplifying an algebraic expression means making it smaller and easier to manage. This section shows you how to simplify. In Chapter 9, you see how important simplifying becomes as you begin solving algebraic equations.

Combining similar terms

When two algebraic terms are *similar* (when their variable parts match), you can add or subtract them (see the earlier section "Considering algebraic terms and the Big Four operations" for details). This feature comes in handy when you're trying to simplify an expression. For example, suppose you're working with following expression:

$$4x - 3y + 2x + y - x + 2y$$

This expression has six terms. But three terms have the variable x and the other three have the variable y. Begin by rearranging the expression to group all similar terms together:

$$= 4x + 2x - x - 3y + y + 2y$$

Now you can add and subtract similar terms. I do this in two steps, first for the x terms and then for the y terms:

$$= 5x - 3y + y + 2y$$
$$= 5x + 0y$$
$$= 5x$$

Notice that the x terms simplify to $5x$ and the y terms simplify to $0y$, which is 0, so the y terms drop out of the expression altogether!

Here's a more complicated example that has variables with exponents:

$$12x - xy - 3x^2 + 8y + 10xy + 3x^2 - 7x$$

This time, you have four different types of terms. Rearrange the terms so that groups of similar terms are all together (I underline these four groups so you can see them clearly):

$$\underline{12x - 7x}\ \underline{-xy + 10xy}\ \underline{-3x^2 + 3x^2}\ \underline{+8y}$$

Now combine each set of similar terms by adding and subtracting:

$$5x + 9xy + 0x^2 + 8y$$

The x^2 terms add up to 0, so they drop out of the expression altogether:

$$= 5x + 9xy + 8y$$

Removing parentheses from an algebraic expression

Parentheses keep parts of an expression together as a single unit. Getting rid of parentheses is often the first step toward solving an algebra problem. In this section, I show how to handle parentheses and the Big Four operations with ease.

Drop everything: Parentheses with a plus sign

When an expression contains parentheses that come right after a plus sign (+), you can just remove the parentheses. Here's an example:

$$2x \underline{+ (3x - y)} + 5y$$
$$= 2x \underline{+ 3x - y} + 5y$$

Now you can simplify the expression by combining similar terms:

$$= 5x + 4y$$

When the first term inside the parentheses is negative, the minus sign replaces the plus sign when you drop the parentheses. For example,

$$6x + (-2x + y) - 4y$$
$$= 6x - 2x + y - 4y$$

Switch signs: Parentheses with a minus sign

Sometimes an expression contains parentheses that come right after a minus sign (–). In this case, change the sign of every term inside the parentheses to the opposite sign; then remove the parentheses.

Consider this example:

$$6x - (2xy - 3y) + 5xy$$

A minus sign is in front of the parentheses, so you need to change the sign of each term in the parentheses and remove the parentheses. Notice that the term $2xy$ appears to have no sign because it's the first term inside the parentheses. This expression really means the following:

$$6x \underline{- (+2xy - 3y)} + 5xy$$

Here's how the signs change:

$$6x \underline{- 2xy + 3y} + 5xy$$

At this point, you can simplify the expression by combining the two similar xy terms:

$$= 6x + 3xy + 3y$$

Distribute: Parentheses with no sign

When you see nothing between a number and a set of parentheses, it means multiplication. For example,

$$2(3) = 6 \qquad 4(4) = 16 \qquad 10(15) = 150$$

This notation becomes much more common with algebraic expressions, replacing even the dot multiplication sign (·):

$$3(4x) = 12x \qquad 4x(2x) = 8x^2 \qquad 3x(7y) = 21xy$$

To remove parentheses without a sign, multiply the term outside the parentheses by every term inside the parentheses; then remove the parentheses.

Here's an example:

$$2(3x - 5y + 4)$$

In this case, multiply 2 by each of the three terms inside the parentheses:

$$= 2(3x) + 2(-5y) + 2(4)$$

For the moment, this expression looks more complex than the original one, but now you can get rid of all three sets of parentheses by multiplying:

$$= 6x - 10y + 8$$

Multiplying by every term inside the parentheses is simply distribution of multiplication over addition — also called the *distributive property* — which I discuss in Chapter 1.

As another example, suppose you have the following expression:

$$-2x(-3x + y + 6) + 2xy - 5x^2$$

Begin by multiplying $-2x$ by the three terms inside the parentheses:

$$-2x(-3x) - 2x(y) - 2x(6) + 2xy - 5x^2$$

The expression looks worse than when you started, but you can get rid of all the parentheses by multiplying:

$$= 6x^2 - 2xy - 12x + 2xy - 5x^2$$

Now you can combine similar terms. I do this in two steps, first rearranging and then combining:

$$= 6x^2 - 5x^2 - 2xy + 2xy - 12x$$
$$= x^2 - 12x$$

FOIL: Two terms in each set of parentheses

Sometimes, expressions have two sets of parentheses next to each other without a sign between them. In that case, you need to multiply every term inside the first set by every term inside the second.

When you have two terms inside each set of parentheses, you can use a process called FOILing. The word *FOIL* is an acronym to help you make sure you multiply the correct terms. It stands for *First, Outside, Inside, and Last.* Here's how the process works:

1. **Start out by multiplying the two *first* terms in the parentheses.**

Suppose you want to simplify the expression $(2x - 2)(3x - 6)$. The first term in the first set of parentheses is $2x$, and the first term in the second set of parentheses is $3x$. Therefore, multiply $2x$ by $3x$:

$$(\underline{2x} - 2)(\underline{3x} - 6)$$
$$2x(3x) = 6x^2$$

2. **Then multiply the two *outside* terms.**

   The two outside terms, $2x$ and -6, are on the ends:

   $$(\underline{2x} - 2)(3x \underline{-6})$$
   $$2x(-6) = -12x$$

3. **Next, multiply the two *inside* terms.**

   The two terms in the middle are -2 and $3x$:

   $$(2x \underline{-2})(\underline{3x} - 6)$$
   $$-2(3x) = -6x$$

4. **Finally, multiply the two *last* terms.**

   The last term in the first set of parentheses is -2, and the last term in the second set is -6:

   $$(2x \underline{-2})(3x \underline{-6})$$
   $$-2(-6) = 12$$

5. **Add these four results together and simplify the expression.**

   $$6x^2 - 12x - 6x + 12$$

   You can further simplify this expression by combining the similar terms $-12x$ and $-6x$:

   $$= 6x^2 - 18x + 12$$

Notice that during this process, you multiply every term inside one set of parentheses by every term inside the other set. FOILing just helps you keep track and make sure you've multiplied everything.

Chapter 9

X's Secret Identity: Solving Algebraic Equations

In This Chapter

- Using variables in equations
- Knowing some quick ways to solve for x in simple equations
- Understanding the balance scale method for solving equations
- Rearranging and isolating terms in an algebraic equation
- Removing parentheses from an equation
- Cross-multiplying to remove fractions

In algebra, solving equations is the main event. Solving an algebraic equation means finding out which number the variable (usually x) stands for. Not surprisingly, this process is called *solving for x,* and when you know how to do it, your confidence in algebra will soar through the roof.

This chapter gives you a bunch of ways to solve equations for the elusive and mysterious x. First, I show you a few informal methods of solving for x. Then I show you how to solve more-difficult equations by thinking of them as a balance scale.

The balance-scale method is really the heart of algebra. After you understand this simple idea, you're ready to solve equations using all the tools I show you in Chapter 8, such as simplifying expressions and removing parentheses. Finally, I show you how cross-multiplying can make solving algebraic equations with fractions a piece of cake.

Understanding Algebraic Equations

An *algebraic equation* is an equation that includes at least one variable — that is, a letter (such as x) that stands for a number. *Solving* an algebraic equation means finding out which number x stands for.

In this section, I show you the basics of how a variable like x works its way into an equation in the first place. Then I show you a few quick ways to solve for x when an equation isn't too difficult.

Using x in equations

An *equation* is a mathematical statement that contains an equal sign. For example, here's a perfectly good equation:

$$7 \cdot 9 = 63$$

At its heart, a variable (such as x) is nothing more than a placeholder for a number. You're probably used to equations that use other placeholders: One number is purposely left as a blank or is replaced with an underline or a question mark, and you're supposed to fill it in. Usually, this number comes after the equal sign. For example,

$$8 + 2 =$$

$$12 - 3 = \_\_$$

$$14 \div 7 = ?$$

As soon as you're comfortable with addition, subtraction, or whatever, you can switch the equations around a bit:

$$9 + \_\_ = 14$$

$$? \cdot 6 = 18$$

When you stop using underlines and question marks and start using variables such as x to stand for the part of the equation you want you to figure out, bingo! You have an algebra problem:

$4 + 1 = x$

$12 \div x = 3$

$x - 13 = 30$

Four ways to solve algebraic equations

Algebra is strong stuff, and you don't always need it to solve an algebraic equation — just as you don't need to call an exterminator just to kill a bug. Generally speaking, you have four ways to solve algebraic equations:

- Eyeballing them (also called *inspection,* or just looking at the problem to get the answer)
- Rearranging them, using inverse operations when necessary
- Guessing and checking
- Applying algebra

Eyeballing easy equations

You can solve easy problems just by looking at them. For example,

$5 + x = 6$

When you look at this problem, you can see that $x = 1$. When a problem is this easy and you can see the answer, you don't need to take any particular trouble to solve it.

Rearranging slightly harder equations

Sometimes rearranging an equation helps turn it into one that you can solve using addition, subtraction, multiplication, or addition. I use this method throughout most of this book, especially when using formulas, such as for geometry problems and problems involving weights and measures. For example, consider the following equation:

$$6x = 96$$

The answer may not jump out at you, but remember that this problem means

$$6 \cdot x = 96$$

You can rearrange this problem using inverse operations (as I show you in Chapter 1):

$$96 \div 6 = x$$

Now solve the problem by division (long or otherwise) to find that $x = 16$.

Guessing and checking equations

You can solve some equations by guessing an answer and then checking to see whether you're right. For example, suppose you want to solve the following equation:

$$3x + 7 = 19$$

To find out what x equals, start by guessing that $x = 2$. Now check to see whether you're right by substituting 2 for x in the equation:

$3(2) + 7 = 6 + 7 = 13 < 19$ Wrong!

When $x = 2$, the left side of the equation equals 13, not 19. That was a low guess, so try a higher guess, $x = 5$:

$3(5) + 7 = 15 + 7 = 22 > 19$ Wrong!

This time, the guess was high, so try $x = 4$:

$3(4) + 7 = 12 + 7 = 19$ Right!

With only three guesses, you find that $x = 4$.

Applying algebra to more-difficult equations

When an algebraic equation gets hard enough, you find that looking at it and rearranging it just isn't enough to solve it. For example, look at this equation:

$$11x - 13 = 9x + 3$$

You probably can't tell what x equals just by looking at this problem. You also can't solve the equation just by using an inverse operation to rearrange it. And guessing and checking would be very tedious. So here's where algebra comes into play.

Algebra is especially useful because you can follow mathematical rules to find your answer. Throughout the rest of this chapter, I show you how to use the rules of algebra to turn tough problems like this one into problems you can solve.

Checks and Balances: Solving for X

An algebra problem is sometimes too complicated to solve just by eyeballing it or rearranging the equation. For these problems, you need a reliable method for getting the right answer. I call this method the *balancing scale.*

The balancing scale allows you to solve for x — that is, find the number that x stands for — in a step-by-step process that always works. In this section, I show you how to use the balancing scale method to solve algebraic equations.

Striking a balance

The equal sign in any equation means that both sides balance. To keep that equal sign, you have to maintain that balance. In other words, whatever you do to one side of an equation, you have to do to the other.

For example, here's a balanced equation:

$$\frac{1+2=3}{\triangle}$$

If you add 1 to one side of the equation, the scale would go out of balance.

$$\underset{\triangle}{\underline{1+2+1 \neq 3}}$$

But if you add 1 to both sides of the equation, the scale stays balanced:

$$\underset{\triangle}{\underline{1+2+1 = 3+1}}$$

You can add any number to the equation, as long as you do it to both sides. And in math, *any number* means x:

$$1 + 2 + x = 3 + x$$

Remember that x is the same wherever it appears in a single equation or problem.

This idea of changing both sides of an equation equally isn't limited to addition. You can just as easily subtract an x, or even multiply or divide by x, as long as you do the same thing to both sides of the equation:

Subtract: $1 + 2 - x = 3 - x$

Multiply: $(1 + 2)x = 3x$

Divide: $\frac{1+2}{x} = \frac{3}{x}$

Using the balance scale to isolate x

The simple idea of balance is at the heart of algebra, and it lets you find out what x is in many equations. When you solve an algebraic equation, the goal is to *isolate* x — that is, to get x alone on one side of the equation and some number on the other side. In algebraic equations of moderate difficulty, this is a three-step process:

1. **Get all constants (non-x terms) on one side of the equation.**

2. **Get all x terms on the other side of the equation.**
3. **Divide to isolate x.**

For example, take a look at the following problem:

$$11x - 13 = 9x + 3$$

As you follow the steps, notice how I keep the equation balanced at each step:

1. **Get all the constants on one side of the equation.**

   Add 13 to both sides of the equation:

$$\begin{array}{l} 11x - 13 = 9x + 3 \\ \underline{\quad +13 \qquad +13} \\ 11x \qquad = 9x + 16 \end{array}$$

   Because you've obeyed the rules of the balance scale, you know that this new equation is also correct. And now, the only non-x term (16) is on the right side of the equation.

2. **Get all the x terms on the other side.**

   Subtract 9x from both sides of the equation:

$$\begin{array}{l} 11x = 9x + 16 \\ \underline{-\ 9x - 9x \qquad} \\ \quad 2x = \qquad 16 \end{array}$$

   Again, the balance is preserved, so the new equation is correct.

3. **Divide by 2 to isolate x.**

$$\frac{2x}{2} = \frac{16}{2}$$

$$x = 8$$

   To check this answer, you can simply substitute 8 for x in the original equation:

$$11(8) - 13 = 9(8) + 3$$

$$88 - 13 = 72 + 3$$

$$75 = 75 \checkmark$$

   This checks out, so 8 is the correct value of x.

Rearranging Equations to Isolate X

The basic tactic to solving algebraic equations is always the same: Changing both sides of the equation equally at every step, you try to isolate x (or whichever variable you're solving for) on one side of the equation.

In this section, I show you how to you put your skills from Chapter 8 to work to solve equations. First, I show you how rearranging the terms in an expression is similar to rearranging them in an equation. Next, I show you how removing parentheses from an equation can help you solve it. Finally, you discover how cross-multiplication is useful for solving algebraic equations that include fractions.

Rearranging terms on one side of an equation

Rearranging terms becomes all-important when working with equations. For example, suppose you're working with this equation:

$$5x - 4 = 2x + 2$$

This equation is really two expressions connected with an equal sign. And of course, that's true of every equation. That's why everything you find out about expressions in Chapter 8 is useful for solving equations. For example, you can rearrange the terms on one side of an equation. So here's another way to write the same equation:

$$-4 + 5x = 2x + 2$$

And here's a third way:

$$-4 + 5x = 2 + 2x$$

This flexibility to rearrange terms comes in handy when you're solving equations.

Moving terms to the other side of the equal sign

Figure 9-1 shows how an equation resembles a balance scale (for more on this idea, see the earlier section "Striking a balance").

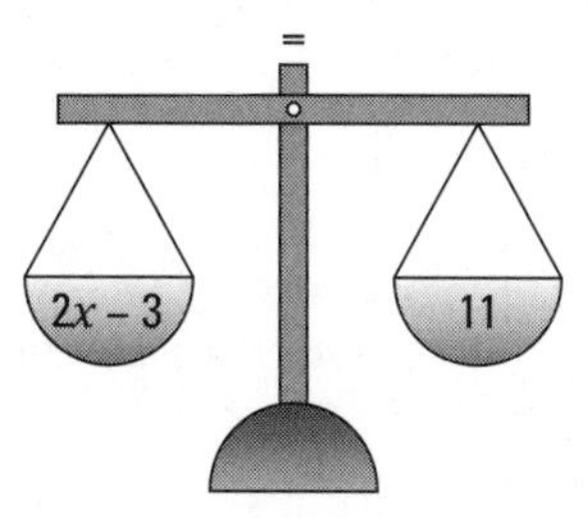

Figure 9-1: Showing how an equation is similar to a balance scale.

If you add or remove anything on one side, you must do the same thing on the other side to keep the scale balanced. For example,

$$\begin{array}{r} 2x-3=11 \\ \underline{-2x \qquad\quad -2x} \\ -3=11-2x \end{array}$$

Now look at these two versions of this equation side by side:

$$2x - 3 = 11 \qquad -3 = 11 - 2x$$

In the first version, the term $2x$ is on the left side of the equal sign. In the second, the term $-2x$ is on the right side. This example illustrates an important rule.

When you move any term in an expression to the other side of the equal sign, change its sign (from plus to minus or from minus to plus).

Suppose you're working with this equation:

$$4x - 2 = 3x + 1$$

You have x's on both sides of the equation, so say you want to move the $3x$. When you move the term $3x$ from the right side to the left side, you have to change its sign from plus to minus (technically, you're just subtracting $3x$ from both sides of the equation):

$$4x - 2 - 3x = 1$$

After that, you can simplify the expression on the left side of the equation by combining similar terms:

$$x - 2 = 1$$

At this point, you can probably see that $x = 3$ because $3 - 2 = 1$. But just to be sure, move the –2 term to the right side and change its sign:

$$x = 1 + 2$$
$$x = 3$$

To check this result, substitute a 3 wherever x appears in the original equation:

$$4(3) - 2 = 3(3) + 1$$
$$12 - 2 = 9 + 1$$
$$10 = 10 \checkmark$$

As you can see, moving terms from one side of an equation to the other can be a big help when you're solving equations.

Removing parentheses from equations

Simplifying expressions (which I discuss in Chapter 8) comes in very handy when you're solving equations. One key skill is removing parentheses from expressions.

For example, suppose you have the following equation:

$$5x + (6x - 15) = 30 - (x - 7) + 8$$

Your mission is to get all the x terms on one side of the equation and all the constants on the other. As the equation stands, however, x terms and constants are "locked together" inside parentheses. Before you can put the x terms on one side of the equation and the constants on the other, you need to remove the parentheses.

An equation is really just two expressions connected by an equal sign, so you can simplify the expressions separately. Start with the expression on the left side. In this expression, the parentheses begin with a plus sign (+), so you can just remove them:

$$5x + \underline{6x - 15} = 30 - (x - 7) + 8$$

Now go to the expression on the right side. This time, the parentheses come right after a minus sign (–). To remove them, change the signs of both terms inside the parentheses: x becomes $-x$, and –7 becomes 7:

$$5x + 6x - 15 = 30 \underline{- x + 7} + 8$$

Bravo! Now you can isolate x terms to your heart's content. Move the $-x$ from the right side of the equation to the left, changing it to x:

$$5x + 6x - 15 \underline{+ x} = 30 + 7 + 8$$

Next, move –15 from the left side to the right, changing it to 15:

$$5x + 6x + x = 30 + 7 + 8 \underline{+ 15}$$

Now combine similar terms on both sides of the equation:

$$12x = 30 + 7 + 8 + 15$$
$$12x = 60$$

Finally, get rid of the coefficient 12 by dividing:

$$\frac{12x}{12} = \frac{60}{12}$$
$$x = 5$$

You can then check your answer by substituting 5 into the original equation wherever x appears:

$$\begin{aligned} 5x+(6x-15) &= 30-(x-7)+8 \\ 5(5)+[6(5)-15] &= 30-(5-7)+8 \\ 25+(30-15) &= 30-(-2)+8 \\ 25+15 &= 30+2+8 \\ 40 &= 40 \checkmark \end{aligned}$$

Here's an example that involves multiplication:

$$11 + 3(-3x + 1) = 25 - (7x - 3) - 12$$

Start out by removing both sets of parentheses. This time, you have no sign between 3 and $(-3x + 1)$ on the left side of the equation, so to remove the parentheses, multiply 3 by both terms inside the parentheses (as I show you in Chapter 8):

$$11 \underline{- 9x + 3} = 25 - (7x - 3) - 12$$

On the right side, the parentheses begin with a minus sign, so remove the parentheses by changing all the signs inside the parentheses:

$$11 - 9x + 3 = 25 \underline{- 7x + 3} - 12$$

Now you're ready to isolate the x terms. I do this in one step, but you can take as many steps as you like:

$$-9x + 7x = 25 + 3 - 12 - 11 - 3$$

At this point, you can combine similar terms:

$$-2x = 2$$

To finish, divide both sides by –2:

$$x = -1$$

Finally, check your solution using substitution:

$$\begin{aligned} 11+3(-3x+1) &= 25-(7x-3)-12 \\ 11+3[\underline{-3(-1)}+1] &= 25-[\underline{7(-1)}-3]-12 \end{aligned}$$

All the variables are gone now, so remember the order of operations rules (from Chapter 2). Begin with the multiplication inside the parentheses, which I've underlined:

$$11 + 3(3 + 1) = 25 - (-7 - 3) - 12$$

Now you can simplify what's inside each set of parentheses:

$$11 + 3(4) = 25 - (-10) - 12$$

Then you can remove the parentheses and complete the check:

$$\begin{aligned} 11 + 12 &= 25 + 10 - 12 \\ 23 &= 23 \checkmark \end{aligned}$$

Using cross-multiplication to remove fractions

In algebra, cross-multiplication helps simplify equations by removing unwanted fractions — and when are fractions ever wanted?

Suppose you want to solve this algebra equation:

$$\frac{x}{2x-2} = \frac{2x+3}{4x}$$

This equation looks hairy. You can't do the division or cancel anything out because the fraction on the left has two terms in the denominator, and the fraction on the right has two terms in the numerator (see Chapter 8 for info on dividing algebraic terms). However, an important piece of information that you have is that the fraction $\frac{x}{2x-2}$ equals the fraction $\frac{2x+3}{4x}$.

When you cross-multiply two equal fractions, you get answers that are equal. For example, here are two equal fractions:

$$\frac{2}{4} = \frac{3}{6}$$

When you *cross-multiply* fractions, you multiply the numerator of one fraction by the denominator of the other:

$$2(6) = 3(4)$$
$$12 = 12 \checkmark$$

So if you cross-multiply the two equal fractions in an algebraic equation, you get two results that are also equal:

$$x(4x) = (2x + 3)(2x - 2)$$

At this point, you have something you know how to work with. The left side is easy:

$$4x^2 = (2x + 3)(2x - 2)$$

The right side requires a bit of FOILing (flip to Chapter 8 for details on how to FOIL):

$$4x^2 = 4x^2 - 4x + 6x - 6$$

Now all the parentheses are gone, so you can isolate the x terms. Because most of these terms are already on the right side of the equation, isolate them on that side:

$$6 = 4x^2 - 4x + 6x - 4x^2$$

Combining similar terms gives you a very pleasant surprise — the two x^2 terms cancel each other out:

$$6 = 2x$$

You may be able to eyeball the correct answer, but here's how to finish:

$$\frac{6}{2} = \frac{2x}{2}$$
$$x = 3$$

Chapter 10

Decoding Algebra Word Problems

In This Chapter

- Solving algebra word problems in simple steps
- Working through solving an algebra word problem
- Choosing variables
- Using charts

Word problems that require algebra are among the toughest problems that students face — and the most common. Teachers just love algebra word problems because they bring together a lot of what you know, such as solving algebra equations (Chapters 8 and 9) and turning words into numbers (Chapters 3 and 7). Standardized tests virtually always include these types of problems.

In this chapter, I show you a five-step method for using algebra to solve word problems. Then I give you a bunch of examples that take you through all five steps.

Along the way, I give you some tips that make solving word problems easier. First, I show you how to choose a variable that makes your equation as simple as possible. Next, I give you practice organizing information from the problem into a chart. By the end of this chapter, you should have a solid understanding of how to solve a wide variety of algebra word problems.

Using a Five-Step Approach

Here are five steps for solving most algebra word problems:

1. **Declare a variable.**
2. **Set up the equation.**
3. **Solve the equation.**
4. **Answer the question that the problem asks.**
5. **Check your answer.**

In this section, I show you how these steps work, using the following word problem as an example:

> In three days, Alexandra sold a total of 31 tickets to her school play. On Tuesday, she sold twice as many tickets as on Wednesday. And on Thursday, she sold exactly 7 tickets. How many tickets did Alexandra sell on each day, Tuesday through Thursday?

Organizing the information in an algebra word problem using a chart or picture is usually helpful. Here's what I came up with:

Tuesday:	twice as many as on Wednesday
Wednesday:	?
Thursday:	7
Total:	31

All the information is in the chart, but the answer still may not be jumping out at you. At this point, you're ready to use the five-step approach to solving the problem.

Note: Everything from Chapters 8 and 9 comes into play when you use algebra to solve word problems, so if you feel a little shaky on solving algebraic equations, flip back to those chapters for review.

Declaring a variable

A *variable* is a letter that stands for a number, and most of the time, you don't find the variable x (or any other variable, for that matter) in a word problem. That doesn't mean you don't need algebra to solve the problem. It just means that you're going to have to put x into the problem yourself and decide what it stands for.

When you *declare a variable,* you say what that variable means in the problem you're solving.

Here are some examples of variable declarations:

Let m = the number of dead mice that the cat dragged into the house.

Let p = the number of times Marianne's husband promised to take out the garbage.

Let c = the number of complaints Arnold received after he painted his garage door purple.

In each case, you take a variable (m, p, or c) and give it a meaning by attaching it to a number.

Notice that the earlier chart for the sample problem has a big question mark next to *Wednesday.* This question mark stands for some number, so you may want to declare a variable that stands for this number. Here's how you do it:

Let w = the number of tickets that Alexandra sold on Wednesday.

Whenever possible, choose a variable with the same initial as what the variable stands for. This practice makes remembering what the variable means a lot easier, which helps you later in the problem.

For the rest of the problem, every time you see the variable w, keep in mind that it stands for the number of tickets that Alexandra sold on Wednesday.

Setting up the equation

After you have a variable to work with, you can go through the problem again and find other ways to use this variable. For example, Alexandra sold twice as many tickets on

Tuesday as on Wednesday, so she sold $2w$ tickets on Tuesday. Now you have a lot more information to fill in the chart:

Tuesday:	Twice as many as on Wednesday	$2w$
Wednesday:	?	w
Thursday:	7	7
Total:	31	31

You know that the total number of tickets — or the sum of the tickets she sold on Tuesday, Wednesday, and Thursday — is 31. With the chart filled in like that, you're ready to set up an equation to solve the problem:

$$2w + w + 7 = 31$$

Solving the equation

After you set up an equation, you can use the tricks from Chapter 9 to solve the equation for w. Here's the equation one more time:

$$2w + w + 7 = 31$$

For starters, remember that $2w$ really means $w + w$. So on the left, you know you really have $w + w + w$, or $3w$; you can simplify the equation a little bit as follows:

$$3w + 7 = 31$$

The goal at this point is to try to get all the terms with w on one side of the equation and all the terms without w on the other side. So on the left side of the equation, you want to get rid of the 7. The inverse of addition is subtraction, so subtract 7 from both sides:

$$\begin{aligned} 3w + 7 &= 31 \\ -7 &\quad -7 \\ \hline 3w \qquad &= 24 \end{aligned}$$

You now want to isolate w on the left side of the equation. To do this, you have to undo the multiplication by 3, so divide both sides by 3:

$$\frac{3w}{3} = \frac{24}{3}$$
$$w = 8$$

Answering the question

You may be tempted to think that after you've solved the equation, you're done. But you still have a bit more work to do. Look back at the word problem, and you see that it asks you this question:

> How many tickets did Alexandra sell on each day, Tuesday through Thursday?

At this point, you have some information that can help you solve the problem. The problem tells you that Alexandra sold 7 tickets on Thursday. Because $w = 8$, you now know that she sold 8 tickets on Wednesday. And on Tuesday, she sold twice as many on Wednesday, so she sold 16. So Alexandra sold 16 tickets on Tuesday, 8 on Wednesday, and 7 on Thursday.

Checking your work

To check your work, compare your answer to the problem, line by line, to make sure every statement in the problem is true:

> In three days, Alexandra sold a total of 31 tickets to her school play.

That's correct, because $16 + 8 + 7 = 31$.

> On Tuesday, she sold twice as many tickets as on Wednesday.

Correct, because she sold 16 tickets on Tuesday and 8 on Wednesday.

> And on Thursday, she sold exactly 7 tickets.

Yep, that's right, too, so you're good to go.

Choosing Your Variable Wisely

When using algebra to solve word problems, you can make the rest of your work a lot easier if you choose your variable wisely. Whenever possible, choose a variable so that the equation you have to solve has no fractions, which are much more difficult to work with than whole numbers.

For example, suppose you're trying to solve this problem:

> Irina has three times as many clients as Toby. If they have 52 clients altogether, how many clients does each person have?

The key sentence in the problem is "Irina has *three times as many* clients as Toby." It's significant because it indicates a relationship between Irina and Toby that's based on either multiplication or division. To avoid fractions, you want to avoid division wherever possible.

Whenever you see a sentence that indicates you should use either multiplication or division, choose your variable to represent the *smaller* number.

In this case, Toby has fewer clients than Irina, so choosing t as your variable is the smart move. Suppose you begin by declaring your variable as follows:

> Let t = the number of clients that Toby has.

Then, using that variable, you can make this chart:

Irina	$3t$
Toby	t

No fraction! Now, to solve this problem, set up this equation:

> Irina + Toby = 52

Plug in the values from the chart:

$$3t + t = 52$$

Now you can solve the problem easily using algebra (as I show you in Chapter 9):

$$4t = 52$$
$$t = 13$$

Toby has 13 clients, so Irina has 39. To check this result — which I highly recommend — note that 13 + 39 = 52.

Now suppose that instead, you take the opposite route and decide to declare a variable as follows:

Let i = the number of clients that Irina has.

Given that variable, you'd have to represent Toby's clients using the fraction $\frac{i}{3}$, which leads to the same answer but a lot more work.

Solving More-Complex Algebra Problems

Algebra word problems become more complex when the number of people or things that you need to find out increases. In this section, the complexity increases from two or three people to four and then five. When you're done, you should feel comfortable solving algebra word problems of significant difficulty.

A chart can help you organize information from a complex word problem so you don't get confused. To see how useful a chart can be, look at this problem, which involves four people:

> Alison, Jeremy, Liz, and Raymond participated in a canned goods drive at work. Liz donated three times as many cans as Jeremy, Alison donated twice as many as Jeremy, and Raymond donated 7 more than Liz. Together, the two women donated two more cans than the two men. How many cans did the four people donate altogether?

The first step, as always, is declaring a variable. To avoid fractions, you want to declare a variable based on the person who brought in the fewest cans. Liz donated more cans than Jeremy, and so did Alison. Furthermore, Raymond donated more cans than Liz. Because Jeremy donated the fewest cans, declare your variable as follows:

Let j = the number of cans that Jeremy donated.

Now you can set up your chart as follows:

Jeremy	j
Liz	$3j$
Alison	$2j$
Raymond	Liz + 7 = $3j + 7$

This looks good because, as expected, there are no fractional amounts in the chart. The next sentence tells you that the women donated two more cans than the men, so make a word equation (as I show you in Chapter 3):

Liz + Alison = Jeremy + Raymond + 2

You can now substitute into this equation as follows, using the information from the chart:

$$3j + 2j = j + 3j + 7 + 2$$

With your equation set up, you're ready to solve. First, isolate the algebraic terms:

$$3j + 2j - j - 3j = 7 + 2$$

Combine similar terms:

$$j = 9$$

Almost without effort, you've solved the equation, so you know that Jeremy donated 9 cans. With this information, you can go back to the chart, plug in 9 for j, and find out how many cans the other people donated: Liz donated 27, Alison donated 18, and Raymond donated 34. Finally, you can add up these numbers to conclude that the four people donated 88 cans altogether.

To check the numbers, read through the problem and make sure they work at every point in the story. For example, together Liz and Alison donated 45 cans and Jeremy and Raymond donated 43, so the women really did donate 2 more cans than the men.

Chapter 11

Geometry: Perimeter, Area, Surface Area, and Volume

In This Chapter

- Examining two-dimensional shapes
- Looking at solid geometry
- Finding out how to measure a variety of shapes

Geometry is the mathematics of figures such as squares, circles, triangles, lines, and so forth. Because geometry is the math of physical space, it's one of the most useful areas of math. Geometry comes into play when measuring rooms or walls in your house, the area of a circular garden, the volume of water in a pool, or the shortest distance across a rectangular field.

In this chapter, I give you the basics on geometric shapes, from flat circles to solid cubes. Then I discuss how to measure geometric shapes by finding the area and perimeter of two-dimensional forms and the volume and surface area of some geometric solids.

Closed Encounters: Understanding 2-D Shapes

Plane geometry is the study of figures on a two-dimensional (2-D) surface — that is, on a plane. You can think of the plane as a piece of paper with no thickness at all. Technically, a plane doesn't end at the edge of the paper — it continues forever.

A *shape* is any closed two-dimensional geometrical figure that has an inside and an outside, separated by the *perimeter* (boundary) of the shape. The *area* of a shape is the measurement of the size inside that boundary.

Measuring the perimeter and area of shapes is useful for a variety of applications, from land surveying (to get information about a parcel of land) to sewing (to figure out how much material you need for a project). In this section, I introduce you to a variety of geometric shapes. Later in the chapter, I show you how to find the perimeter and area of each, but for now I just acquaint you with the shapes.

Circles

A *circle* is the set of all points that are a constant distance from the circle's center. The distance from any point on the circle to its center is called the *radius* of the circle. The distance from any point on the circle straight through the center to the other side of the circle is called the *diameter* of the circle.

Polygons

A *polygon* is any shape whose sides are all straight. Every polygon has three or more sides (if it had fewer than three, it wouldn't really be a shape at all). Following are some of the most common polygons:

- **Triangles:** The most basic shape with straight sides is the *triangle,* a three-sided polygon. You find out all about triangles when you study trigonometry (and what better place to begin than *Trigonometry For Dummies*?). Triangles are classified on the basis of their sides (equilateral, isosceles, or scalene) and angles (right, acute, or obtuse).
- **Quadrilaterals:** A *quadrilateral* is any shape that has four straight sides. Quadrilaterals are some of the most common shapes you see in daily life. If you doubt this, look around and notice that most rooms, doors, windows, and tabletops are quadrilaterals. Some basic quadrilaterals include the square, rectangle, rhombus, parallelogram, trapezoid, and kite.

A polygon can have any number of sides. Polygons with more than four sides are not as common as triangles and quadrilaterals, but they're still worth knowing about.

Larger polygons come in two basic varieties: regular and irregular. A *regular polygon* has equal sides and equal angles; the most common are regular pentagons (five sides), regular hexagons (six sides), and regular octagons (eight sides). Every other polygon is an *irregular polygon.*

Adding Another Dimension: Solid Geometry

Solid geometry is the study of shapes in *space* — that is, the study of shapes in three dimensions. A *solid* is the spatial (three-dimensional, or 3-D) equivalent of a shape. Every solid has an inside and an outside separated by the *surface* of the solid. Here, I introduce you to a variety of solids.

The many faces of polyhedrons

A polyhedron is the three-dimensional equivalent of a polygon, a shape that has only straight sides. Similarly, a *polyhedron* is a solid that has only straight edges and flat faces (that is, faces that are polygons). Figure 11-1 shows some examples of common polyhedrons.

The most common polyhedron is the *cube.* A cube has six flat faces that are polygons — in this case, squares — and 12 straight edges. Additionally, a cube has eight *vertexes* (corners).

Later in this chapter, I show you how measure each of the common polyhedrons to determine its *volume* — that is, the amount of space contained inside its surface.

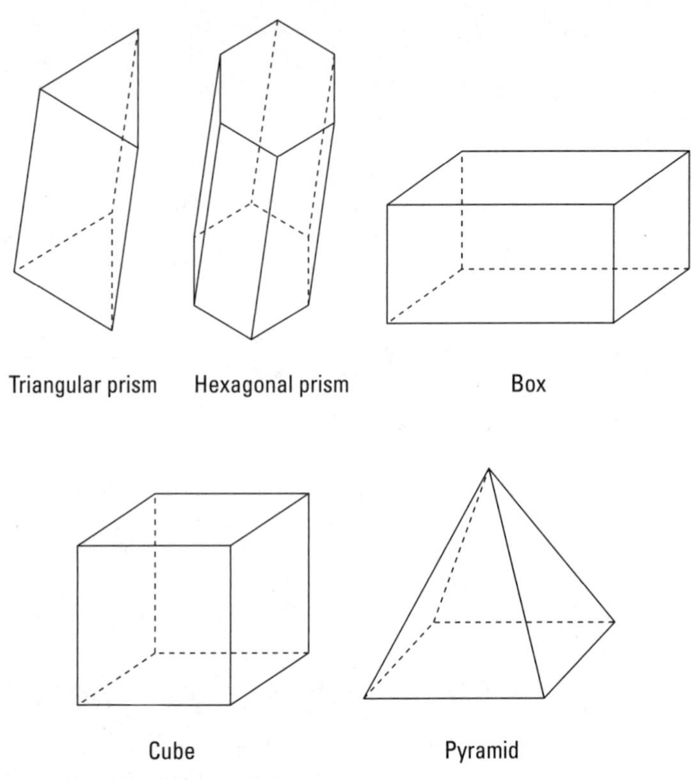

Figure 11-1: Common polyhedrons.

3-D shapes with curves

Many solids aren't polyhedrons because they contain at least one curved surface. Here are a few of the most common of these types of solids (see Figure 11-2):

- **Sphere:** A *sphere* is the solid, or three-dimensional, equivalent of a circle. A ball is a perfect visual aid for a sphere.
- **Cylinder:** A *cylinder* has a circular base and extends vertically from the plane. A good visual aid for a cylinder is a can of soup.
- **Cone:** A *cone* is a solid with a round base that extends vertically to a single point. A good visual aid for a cone is an ice cream cone.

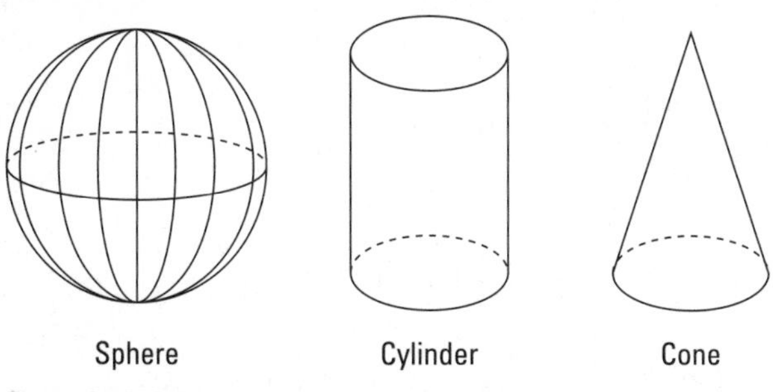

Figure 11-2: Spheres, cylinders, and cones.

In the next section, I show you how to measure each of these solids to determine its volume.

Measuring Shapes: Perimeter, Area, Surface Area, and Volume

In this section, I introduce you to some important formulas for measuring shapes and solids. These formulas use letters, or *variables,* to stand for numbers that you can plug in to make specific measurements.

2-D: Measuring on the flat

Two important skills in geometry — and real life — are finding the perimeter and the area of shapes. A shape's *perimeter* is a measurement of the length of its sides. You use perimeter for measuring the distance around the edges of a room, building, or circular pathway. A shape's *area* is a measurement of how big its surface is. You use area when measuring the size of a wall, a table, or a pizza.

In Figure 11-3, I give you the lengths of the sides of a few shapes.

When every side of a shape is straight, you can measure its perimeter by adding up the lengths of all of its sides.

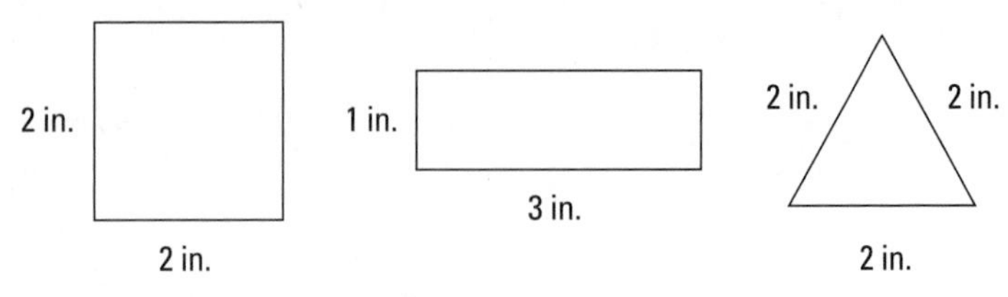

Figure 11-3: Measuring the sides of shapes.

Similarly, in Figure 11-4, I give you the area of each shape.

The area of a shape is always measured in square units: square inches (in.^{2}), square feet (ft.^{2}), square miles (mi.^{2}), square kilometers (km^2), and so on — even if you're talking about the area of a circle.

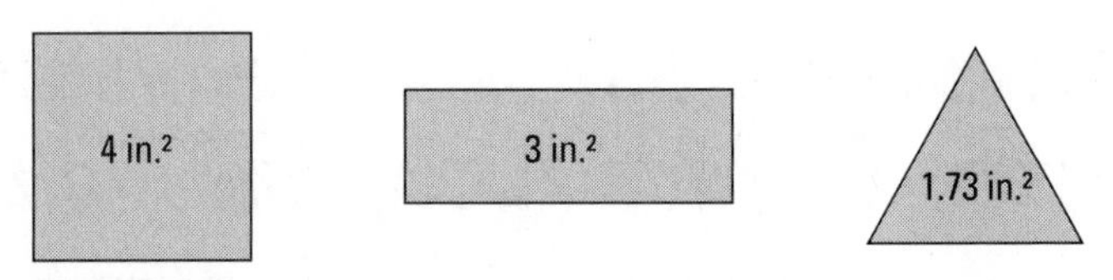

Figure 11-4: The areas of shapes.

Going 'round in circles

The *center* of a circle is a point that's the same distance from any point on the circle itself. This distance is called the *radius* of the circle, or *r* for short. And any line segment from one point on the circle through the center to another point on the circle is called a *diameter,* or *d* for short. See Figure 11-5.

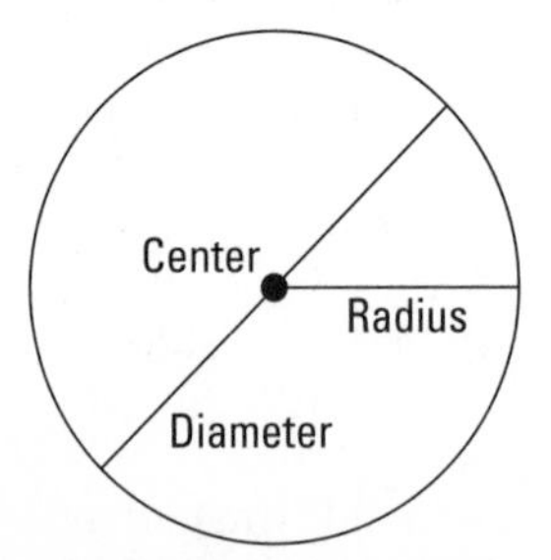

Figure 11-5: The parts of a circle.

As you can see, the diameter of any circle is made up of one radius plus another radius — that is, two radii. This concept gives you the following handy formula:

$$d = 2 \cdot r$$

For example, given a circle with a radius of 5 mm, you can figure out the diameter as follows:

$$d = 2 \cdot 5 \text{ mm} = 10 \text{ mm}$$

Because the circle is a special shape, its perimeter (the length of its "sides") has a special name: the *circumference* (C for short). Early mathematicians went to a lot of trouble figuring out how to measure the circumference of a circle. Here's the formula they hit upon:

$$C = 2 \cdot \pi \cdot r$$

Note: Because $2 \cdot r$ is the same as the diameter, you also can write the formula as $C = \pi \cdot d$

The symbol π is called *pi.* It's just a number whose approximate value is as follows (the decimal part of pi goes on forever, so you can't get an exact value for pi):

$$\pi \approx 3.14$$

So given a circle with a radius of 5 mm, you can figure out the approximate circumference:

$$C \approx 2 \cdot 3.14 \cdot 5 \text{ mm} = 31.4 \text{ mm}$$

The formula for the area *(A)* of a circle also uses π:

$$A = \pi \cdot r^2$$

Here's how to use this formula to find the approximate area of a circle with a radius of 5 mm:

$$A \approx 3.14 \cdot (5 \text{ mm})^2 = 3.14 \cdot 25 \text{ mm}^2 = 78.5 \text{ mm}^2$$

Measuring triangles

In this section, I discuss how to measure the perimeter and area of all triangles. Then I show you a special feature of right triangles that allows you to measure them more easily.

Finding the perimeter and area of a triangle

Mathematicians have no special formula for finding the perimeter of a triangle — they just add up the lengths of the sides.

To find the area of a triangle, you need to know the length of one side — the *base* (*b* for short) — and the height *(h)*. Note that the *height* forms a right angle with the base. Figure 11-6 shows a triangle with a base of 5 cm and a height of 2 cm.

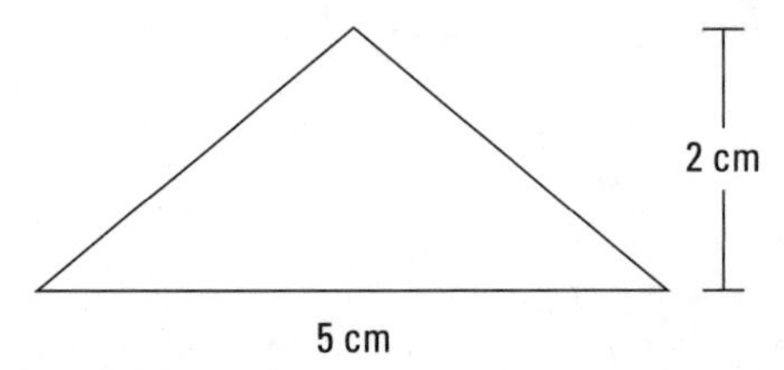

Figure 11-6: The base and height of a triangle.

Here's the formula for the area of a triangle:

$$A = \frac{1}{2}(b \cdot h)$$

So here's how to figure out the area of a triangle with a base of 5 cm and a height of 2 cm:

$$A = \frac{1}{2}(5 \text{ cm} \cdot 2 \text{ cm}) = \frac{1}{2}\left(10 \text{ cm}^2\right) = 5 \text{ cm}^2$$

Lessons from Pythagoras: Finding the third side of a right triangle

A *right triangle* has one right (90°) angle. The long side of a right triangle *(c)* is called the *hypotenuse,* and the two short sides *(a* and *b)* are called the *legs* (see Figure 11-7). The most important right triangle formula is the *Pythagorean theorem:*

$$a^2 + b^2 = c^2$$

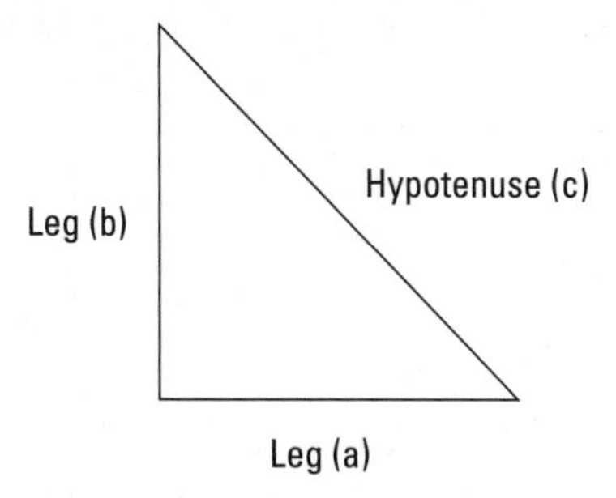

Figure 11-7: The hypotenuse and legs of a right triangle.

This formula allows you to find the hypotenuse of a triangle given only the lengths of the legs. For example, suppose the legs of a triangle are 3 and 4 units. Here's how to use the Pythagorean theorem to find the length of the hypotenuse:

$$3^2 + 4^2 = c^2$$
$$9 + 16 = c^2$$
$$25 = c^2$$

So when you multiply c by itself, the result is 25. Therefore,

$$c = 5$$

The length of the hypotenuse is 5 units.

Measuring squares

The letter s represents the length of a square's side. For example, if the side of a square is 3 inches, then you say $s = 3$ in. Finding the perimeter *(P)* of a square is simple: Just multiply the length of the side by 4. Here's the formula for the perimeter of a square:

$$P = 4 \cdot s$$

For example, if the length of the side is 3 inches, substitute 3 inches for s in the formula:

$$P = 4 \cdot 3 \text{ in.} = 12 \text{ in.}$$

Finding the area of a square is also easy: Just multiply the length of the side by itself — that is, *take the square* of the side. Here are two ways of writing the formula for the area of a square (s^2 is pronounced *s squared*):

$$A = s^2, \text{ or } A = s \cdot s$$

For example, if the length of the side is 3 inches, then you get the following:

$$A = (3 \text{ in.})^2 = 3 \text{ in.} \cdot 3 \text{ in.} = 9 \text{ in.}^2$$

Working with rectangles

The long side of a rectangle is called the *length,* or l for short. The short side is called the *width,* or w for short. For example, in a rectangle whose sides are 5 and 4 feet long, $l = 5$ ft. and $w = 4$ ft.

Because a rectangle has two lengths and two widths, you can use the following formula for the perimeter of a rectangle:

$$P = 2 \cdot (l + w)$$

Calculate the perimeter of a rectangle whose length is 5 yards and whose width is 4 yards as follows:

$$P = 2 \cdot (5 \text{ yd.} + 4 \text{ yd.}) = 2 \cdot 9 \text{ yd.} = 18 \text{ yd.}$$

The formula for the area of a rectangle is

$$A = l \cdot w$$

So here's how you calculate the area of the same rectangle:

$$A = 5 \text{ yd.} \cdot 4 \text{ yd.} = 20 \text{ yd.}^2$$

Calculating with rhombuses

A *rhombus* is like a square that's been collapsed as if its corners were hinges. All four sides are equal in length, and both pairs of opposite sides are parallel.

As with a square, use s to represent the length of a rhombus's side. But another key measurement for a rhombus is its height. The *height* of a rhombus (h for short) is the shortest distance from one side to the opposite side. In the Figure 11-8, $s = 4$ cm and $h = 2$ cm.

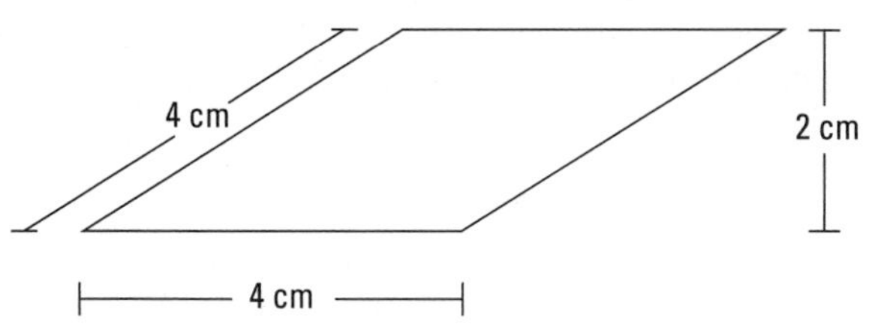

Figure 11-8: Measuring a rhombus.

The formula for the perimeter of a rhombus is the same as for a square:

$$P = 4 \cdot s$$

Here's how you figure out the perimeter of a rhombus whose side is 4 centimeters:

$$P = 4 \cdot 4 \text{ cm} = 16 \text{ cm}$$

To measure the area of a rhombus, you need both the length of a side and the height. Here's the formula:

$$A = s \cdot h$$

So here's how you determine the area of a rhombus with a side of 4 cm and a height of 2 cm:

$$A = 4 \text{ cm} \cdot 2 \text{ cm} = 8 \text{ cm}^2$$

You can read 8 cm^2 as 8 *square centimeters* or, less commonly, as 8 *centimeters squared.*

Measuring parallelograms

Imagine taking a rectangle and collapsing it as if the corners were hinges. This shape is a *parallelogram* — both pairs of opposite sides are equal in length, and both pairs of opposite sides are parallel.

The top and bottom sides of a parallelogram are called its *bases* (b for short), and the remaining two sides are its *sides* (s). And as with rhombuses, another important measurement of a parallelogram is its *height* (h), the shortest distance between the bases. So the parallelogram in Figure 11-9 has these measurements: $b = 6$ in., $s = 3$ in., and $h = 2$ in.

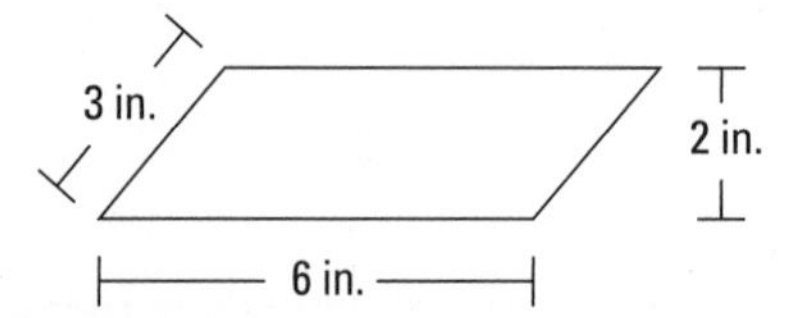

Figure 11-9: Measuring a parallelogram.

Each parallelogram has two equal bases and two equal sides. Therefore, here's the formula for the perimeter of a parallelogram:

$$P = 2 \cdot (b + s)$$

To figure out the perimeter of the parallelogram in this section, just substitute in the measurements for the bases and sides:

$$P = 2 \cdot (6 \text{ in.} + 3 \text{ in.}) = 2 \cdot 9 \text{ in} = 18 \text{ in.}$$

Here's the formula for the area of a parallelogram:

$$A = b \cdot h$$

And here's how you calculate the area of the same parallelogram:

$$A = 6 \text{ in.} \cdot 2 \text{ in.} = 12 \text{ in.}^2$$

Measuring trapezoids

A *trapezoid* has four sides, and at least two opposite sides are parallel. The parallel sides of a trapezoid are called its *bases.* Because these bases are different lengths, you can call them b_1 and b_2. The *height* (h) of a trapezoid is the shortest distance between the bases. Thus, the trapezoid in Figure 11-10 has these measurements: $b_1 = 2$ in., $b_2 = 3$ in., and $h = 2$ in.

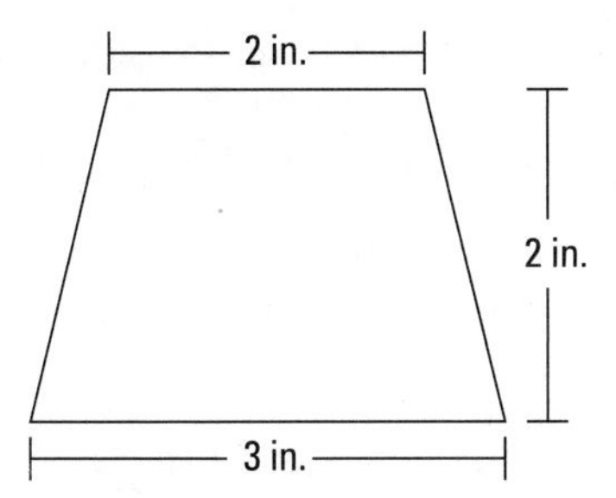

Figure 11-10: Measuring a trapezoid.

Because a trapezoid can have sides of four different lengths, you really don't have a special formula for finding the perimeter of a trapezoid. Just add up the lengths of its sides and you get your answer.

Here's the formula for the area of a trapezoid:

$$A = \frac{1}{2} \cdot (b_1 + b_2) \cdot h$$

So here's how to find the area of the pictured trapezoid:

$$\begin{aligned} A &= \frac{1}{2} \cdot (2 \text{ in.} + 3 \text{ in.}) \cdot 2 \text{ in.} \\ &= \frac{1}{2} \cdot 5 \text{ in.} \cdot 2 \text{ in.} \\ &= \frac{1}{2} \cdot 10 \text{ in.}^2 \\ &= 5 \text{ in.}^2 \end{aligned}$$

Note: Because of the associative property (see Chapter 1), I'm allowed to multiply 5 in. · 2 in. before multiplying by $\frac{1}{2}$.

Spacing out: Measuring in three dimensions

In three dimensions, the concepts of perimeter and area have to be tweaked a little. In 2-D, the *perimeter* of a shape is the measurement of its boundary, and the *area* of a shape is the measurement of the surface inside the boundary. In 3-D, the boundary of a solid is called its *surface area,* and what's inside a solid is called its *volume.*

The *surface area* of a solid is a measurement of the size of its surface, as measured in square units such as square inches (in.^{2}), square feet (ft.^{2}), square meters (m^2), and so forth. The *volume (V)* of a solid is a measurement of the space it occupies, as measured in cubic units such as cubic inches (in.^{3}), cubic feet (ft.^{3}), cubic meters (m^3), and so forth.

You can find the surface area of a *polyhedron* (a solid whose faces are all polygons) by adding together the areas of all its faces. The preceding sections give you some area formulas for common shapes. In most cases, you don't need to know separate formulas for finding the surface area of a solid.

Finding the volume of solids, however, is something mathematicians love to know. In the next subsections, I give you the formulas for finding the volumes of a variety of solids.

Spheres

The *center* of a sphere is a point that's the same distance from any point on the sphere itself. This distance is called the *radius (r)* of the sphere. If you know the radius of a sphere, you can find out its volume using the following formula:

$$V = \frac{4}{3} \cdot \pi \cdot r^3$$

For example, here's how to figure out the approximate volume of a ball whose radius is 4 inches:

$$\begin{aligned} V &\approx \frac{4}{3} \cdot 3.14 \cdot (4 \text{ in.})^3 \\ &\approx \frac{4}{3} \cdot 3.14 \cdot 64 \text{ in.}^3 \\ &\approx 4.19 \cdot 64 \text{ in.}^3 \\ &= 268.16 \text{ in.}^3 \end{aligned}$$

This is the approximate volume because I use 3.14 for π. In the preceding problem, I use equal signs when a value is equal to whatever comes right before it and approximately-equal-to signs ($\approx$) when I round.

Cubes

The main measurement of a cube is the length of its side *(s)*. Using this measurement, you can find out the volume of a cube using the following formula:

$$V = s^3$$

So if the side of a cube is 5 meters, here's how you figure out its volume:

$$V = (5\text{ m})^3 = 5\text{ m} \cdot 5\text{ m} \cdot 5\text{ m} = 125\text{ m}^3$$

You can read 125 m^3 as 125 *cubic meters* or, less commonly, as 125 *meters cubed.*

Boxes (Rectangular solids)

The three measurements of a box (or rectangular solid) are its length *(l)*, width *(w)*, and height *(h)*. The box in Figure 11-11 has the following measurements: $l = 4$ m, $w = 3$ m, and $h = 2$ m.

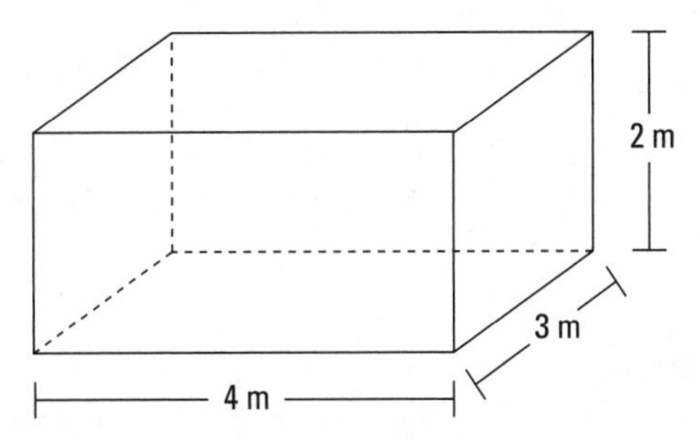

Figure 11-11: Measuring a box.

You can find the volume of a box using the following formula:

$$V = l \cdot w \cdot h$$

So here's how to find the volume of the box pictured in this section:

$$V = 4\text{ m} \cdot 3\text{ m} \cdot 2\text{ m} = 24\text{ m}^3$$

Prisms

Finding the volume of a prism (see prisms in Figure 11-1) is easy if you have two measurements. One measurement is the height *(h)* of the prism. The second is the *area of the base* (A_b). The *base* is the polygon that extends vertically from the plane. (I show you how to find the area of a variety of *polygons,* or shapes with straight sides, earlier in "2-D: Measuring on the flat.")

Here's the formula for finding the volume of a prism:

$$V = A_b \cdot h$$

For example, suppose a prism has a base with an area of 5 square centimeters and a height of 3 centimeters. Here's how you find its volume:

$$V = 5 \text{ cm}^2 \cdot 3 \text{ cm} = 15 \text{ cm}^3$$

Notice that the units of measurements (cm^2 and cm) are also multiplied, giving you a result of cm^3.

Cylinders

You find the volume of cylinders the same way you find the area of prisms — by multiplying the area of the base (A_b) by the cylinder's height *(h)*:

$$V = A_b \cdot h$$

Suppose you want to find the volume of a cylindrical can whose height is 4 inches and whose base is a circle with a radius of 2 inches. First, find the area of the base by using the formula for the area of a circle:

$$\begin{aligned} A_b &= \pi \cdot r^2 \\ &\approx 3.14 \cdot (2 \text{ in.})^2 \\ &\approx 3.14 \cdot 4 \text{ in.}^2 \\ &= 12.56 \text{ in.}^2 \end{aligned}$$

This area is approximate because I use 3.14 as an approximate value for π.

Now use this area to find the volume of the cylinder:

$$V \approx 12.56 \text{ in.}^2 \cdot 4 \text{ in.} = 50.24 \text{ in.}^3$$

Notice how multiplying square inches (in.^{2}) by inches gives a result in cubic inches (in.^{3}).

Pyramids and cones

The two key measurements for pyramids and cones are the same as those for prisms and cylinders (see the preceding sections): the height (h) and the area of the base (A_b). Here's the formula for the volume of a pyramid or a cone:

$$V = \frac{1}{3}(A_b \cdot h)$$

For example, suppose you want to find the volume of an ice cream cone whose height is 4 inches and whose base area is 3 square inches. Here's how you do it:

$$\begin{aligned} V &= \frac{1}{3}(3 \text{ in.}^2 \cdot 4 \text{ in.}) \\ &= \frac{1}{3}(12 \text{ in.}^3) \\ &= 4 \text{ in.}^3 \end{aligned}$$

Similarly, suppose you want to find the volume of a pyramid in Egypt. It has a height of 60 meters and a square base whose sides are each 50 meters. First, find the area of the base using the formula for the area of a square from "2-D: Measuring on the flat," earlier in this chapter:

$$A_b = s^2 = (50 \text{ m})^2 = 2{,}500 \text{ m}^2$$

Now use this area to find the volume of the pyramid:

$$\begin{aligned} V &= \frac{1}{3}(2{,}500 \text{ m}^2 \cdot 60 \text{ m}) \\ &= \frac{1}{3}(150{,}000 \text{ m}^3) \\ &= 50{,}000 \text{ m}^3 \end{aligned}$$

Chapter 12

Picture It! Graphing Information

In This Chapter

- Reading bar graphs, pie charts, and line graphs
- Understanding the Cartesian coordinate system
- Plotting points and lines on a graph
- Solving problems using a graph

A *graph* is a visual tool for organizing and presenting information about numbers. Most students find graphs relatively easy because they provide a picture to work with rather than just a bunch of numbers. Their simplicity makes graphs show up in newspapers, magazines, business reports, and anywhere else clear visual communication is important.

In this chapter, I introduce you to three common styles of graphs: the bar graph, the pie chart, and the line graph. I show you how to read each of these styles of graphs to obtain information. I also show you how to answer the types of questions people may ask to check your understanding.

I spend the rest of the chapter focusing on the most important type of mathematical graph: the *Cartesian coordinate system,* or *x-y* plane. This system is so common that when math folks talk about a graph, they're usually talking about this type. I show you the various parts of the graph, and then I show you how to plot points and lines. At the end of the chapter, you see how you can solve math problems using a graph.

Examining Three Common Graph Styles

In this section, I show you how to read and understand three styles of graphs: the bar graph, the pie chart, and the line graph. They aren't the only types of graphs, but they're very common, and understanding them can give you a leg up on reading other types of graphs when you see them.

Each of these graph styles has a specific function:

- The bar graph is best for representing numbers that are independent of each other.
- The pie chart allows you to show how a whole is cut up into parts.
- The line graph gives you a sense of how numbers change over time.

Bar graph

A *bar graph* gives you an easy way to compare numbers or values. For example, Figure 12-1 shows a bar graph comparing the performance of five trainers at a fitness center.

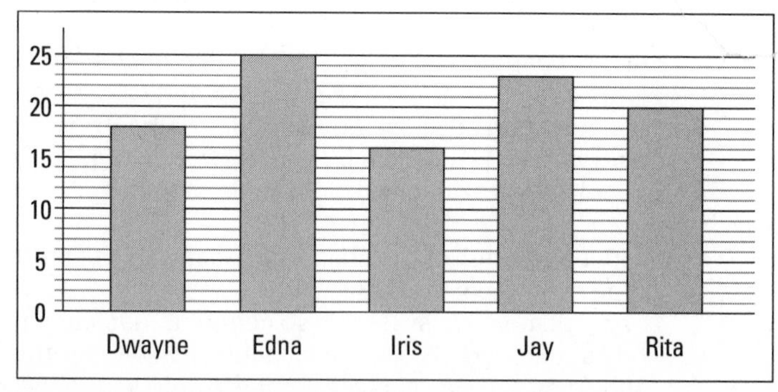

Figure 12-1: The number of new clients recorded this quarter.

As you can see from the caption, the graph shows how many new clients each trainer has enrolled this quarter. The advantage of such a graph is that you can see at a glance, for example, that Edna has the most new clients and Iris has the fewest. The bar graph is a good way to represent numbers that are independent of each other. For example, if Iris gets another new client, it doesn't necessarily affect any other trainer's performance.

Reading a bar graph is easy after you get used to it. Here are a few types of questions someone could ask about the bar graph in Figure 12-1:

- **Individual values:** *How many new clients does Jay have?* Find the bar representing Jay's clients and notice that he has 23 new clients.
- **Differences in value:** *How many more clients does Rita have compared with Dwayne's?* Notice that Rita has 20 new clients and Dwayne has 18, so she has 2 more than he does.
- **Totals:** *Together, how many clients do the three women have?* Notice that the three women — Edna, Iris, and Rita — have 25, 16, and 20 new clients, respectively, so they have 61 new clients altogether.

Pie chart

A *pie chart,* which looks like a divided circle, shows you how a whole object is cut up into parts. Pie charts are most often used to represent percentages. For example, Figure 12-2 shows a pie chart representing Eileen's monthly expenses.

You can tell at a glance that Eileen's largest expense is rent and that her second largest is her car. Unlike the bar graph, the pie chart shows numbers that are dependent upon each other. For example, if Eileen's rent increases to 30% of her monthly income, she'll have to decrease her spending in at least one other area.

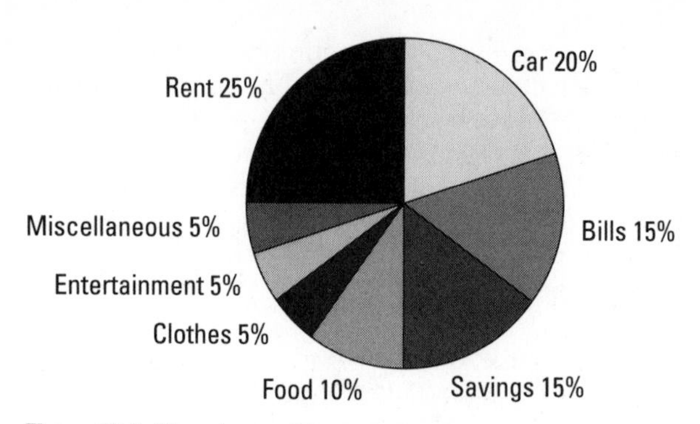

Figure 12-2: Eileen's monthly expenses.

Here are a few typical questions you may be asked about a pie chart:

- **Individual percentages:** *What percentage of her monthly expenses does Eileen spend on food?* Find the slice that represents what Eileen spends on food, and notice that she spends 10% of her income there.
- **Differences in percentages:** *What percentage more does she spend on her car than on entertainment?* Eileen spends 20% on her car but only 5% on entertainment, so the difference between these percentages is 15%.
- **How much a percent represents in terms of dollars:** *If Eileen brings home $2,000 per month, how much does she put away in savings each month?* First notice that Eileen puts 15% into savings every month. So you need to figure out 15% of $2,000. Using your skills from Chapter 6, solve this problem by turning 15% into a decimal and multiplying:

  $$0.15 \cdot 2{,}000 = 300$$

  So Eileen saves $300 every month.

Line graph

The most common use of a *line graph* is to plot how numbers change over time. For example, Figure 12-3 is a line graph showing last year's sales figures for Tami's Interiors.

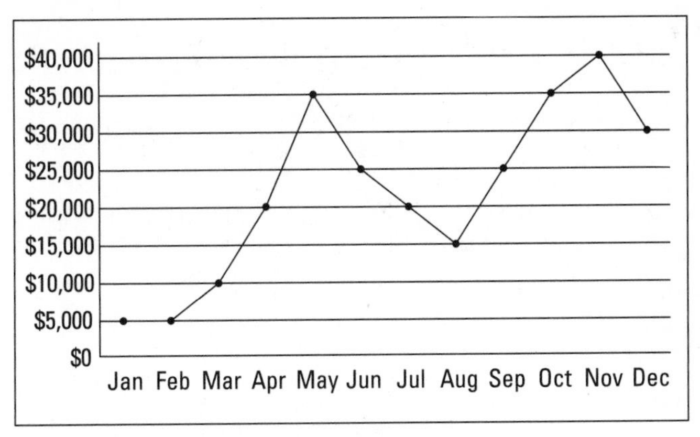

Figure 12-3: Gross receipts for Tami's Interiors.

The line graph shows a progression in time. At a glance, you can tell that Tami's business tended to rise strongly at the beginning of the year, drop off during the summer, rise again in the fall, and then drop off again in December.

Here are a few typical questions that you may be asked to show that you know how to read a line graph:

- **High or low points and timing:** *In what month did Tami bring in the most revenue, and how much did she bring in?* Notice that the highest point on the graph is in November, when Tami's revenue reached $40,000.
- **Total over a period of time:** *How much did she bring in altogether during the last quarter of the year?* A quarter of a year is three months, so the last quarter is the last three months of the year. Tami brought in $35,000 in October, $40,000 in November, and $30,000 in December, so her total receipts for the last quarter add up to $105,000.
- **Greatest change:** *In what month did the business show the greatest gain in revenue as compared with the previous month?* You want to find the line segment on the graph that has the steepest upward slope. This change occurs between April and May, where Tami's revenue increased by $15,000, so her business showed the greatest gain in May.

Using Cartesian Coordinates

When math folks talk about using a graph, they're usually referring to a *Cartesian graph* (also called the Cartesian coordinate system), as Figure 12-4 shows. You see a lot of this graph when you study algebra, so getting familiar with it now is a good idea.

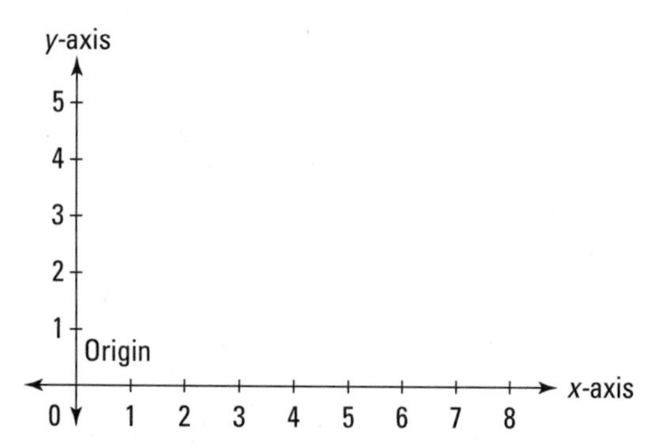

Figure 12-4: A Cartesian graph includes horizontal and vertical axes, which cross at the origin (0, 0).

A Cartesian graph is really just two number lines that cross at 0. These number lines are called the *horizontal axis* (also called the *x-axis*) and the *vertical axis* (also called the *y-axis*). The place where these two axes cross is called the *origin.*

Plotting points on a Cartesian graph

Plotting a point (finding and marking its location) on a graph isn't much harder than finding a point on a number line, because a graph is just two number lines put together.

Every point on a Cartesian graph is represented by a *set of coordinates* — two numbers in parentheses, separated by a comma. To plot any point, start at the origin, where the two axes cross. The first number, or x value, tells you how far to go to the right (if positive) or left (if negative) along the horizontal axis. The second number, or y value, tells you how far to go up (if positive) or down (if negative) along the vertical axis.

For example, here are the coordinates of four points called A, B, C, and D:

$A = (2, 3)$ $B = (-4, 1)$

$C = (0, -5)$ $D = (6, 0)$

Figure 12-5 depicts a graph with these four points plotted. Start at the origin, (0, 0). To plot point A, count 2 spaces to the right and 3 spaces up. To plot point B, count 4 spaces to the left (the negative direction) and then 1 space up. To plot point C, count 0 spaces left or right and then count 5 spaces down (the negative direction). And to plot point D, count 6 spaces to the right and then 0 spaces up or down.

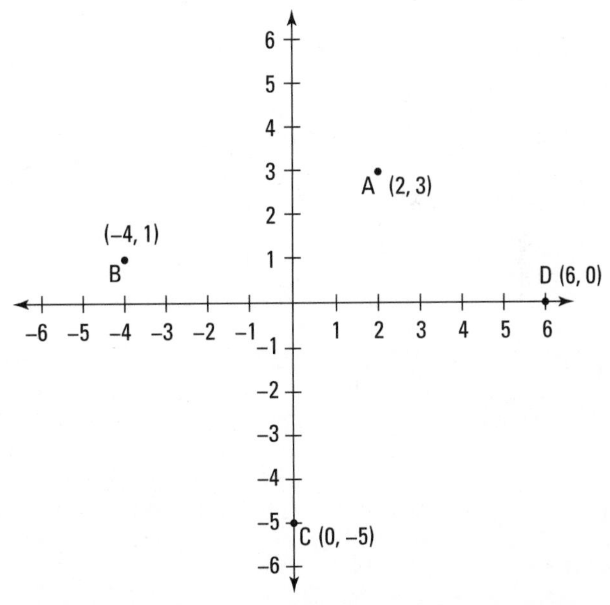

Figure 12-5: Points A, B, C, and D plotted on a Cartesian graph.

Drawing lines on a Cartesian graph

After you understand how to plot points on a graph (see the preceding section), you can begin to plot lines and use them to show mathematical relationships.

The examples in this section focus on the number of dollars that two people, Xenia and Yanni, are carrying. The horizontal axis represents Xenia's money, and the vertical axis represents Yanni's. For example, suppose you want to draw a line representing this statement:

> Xenia has $1 more than Yanni.

To do this, make a chart as follows:

Xenia	1	2	3	4	5
Yanni					

Now fill in each column of the chart, assuming that Xenia has that number of dollars. For example, if Xenia has $1, then Yanni has $0. And if Xenia has $2, then Yanni has $1. Continue until your chart looks like this:

Xenia	1	2	3	4	5
Yanni	0	1	2	3	4

Now you have five pairs of points that you can plot on your graph as (Xenia, Yanni): (1, 0), (2, 1), (3, 2), (4, 3), and (5, 4). Next, draw a straight line through these points, as in Figure 12-6.

This line on the graph represents every possible pair of amounts for Xenia and Yanni. For example, notice how the point (6, 5) is on the line. This point represents the possibility that Xenia has $6 and Yanni has $5.

Here's a slightly more complicated example:

> Yanni has $3 more than twice the amount that Xenia has.

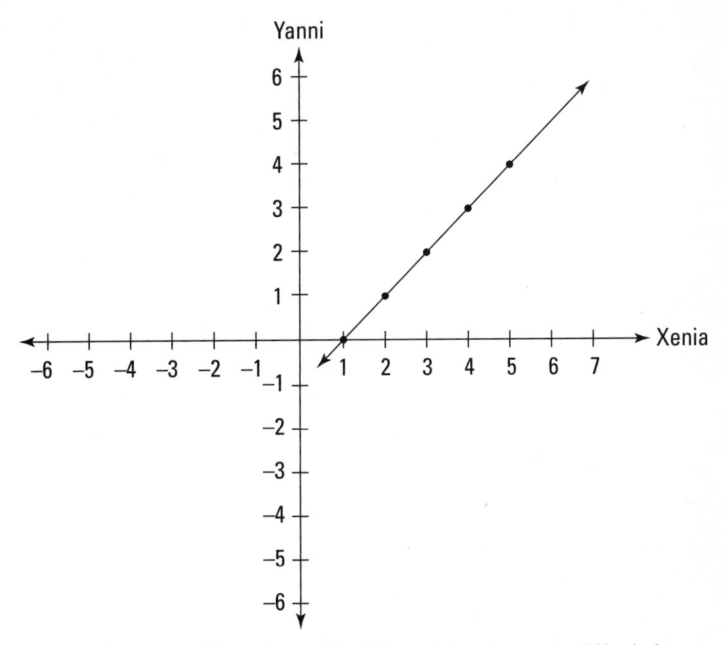

Figure 12-6: All possible values of Xenia's and Yanni's money if Xenia has $1 more than Yanni.

Again, start by making the usual chart:

Xenia	1	2	3	4	5
Yanni					

You can fill in this chart by supposing that Xenia has a certain amount of money and then figuring out how much money Yanni would have in that case. For example, if Xenia has $1, then twice that amount is $2; Yanni has $3 more than that, or $5. And if Xenia has $2, then twice that amount is $4; Yanni has $3 more, or $7. Continue in that way to fill in the chart as follows:

Xenia	1	2	3	4	5
Yanni	5	7	9	11	13

Now plot these five points on the graph and draw a line through them, as Figure 12-7 shows. As in the other examples, this graph represents all possible values that Xenia and Yanni could have. For example, if Xenia has $7, Yanni has $17.

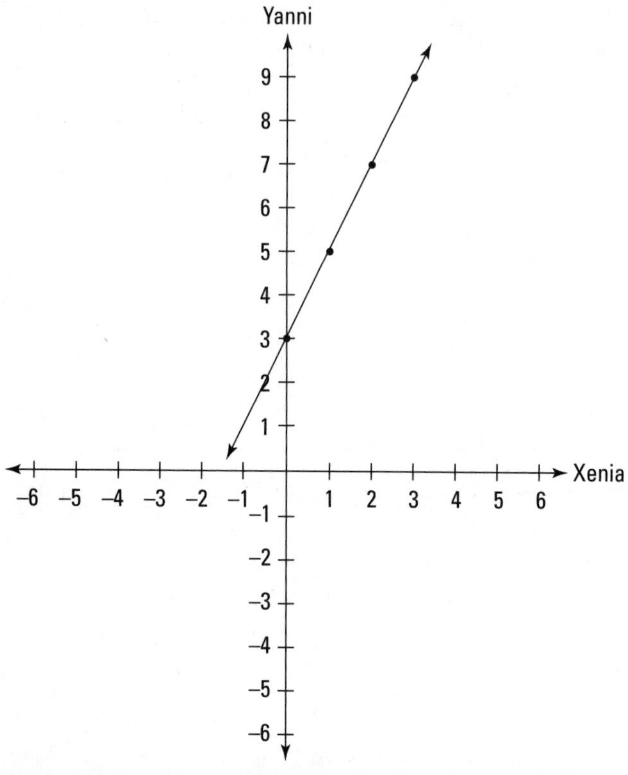

Figure 12-7: Possible values of Xenia's and Yanni's money if Yanni has $3 more than twice the amount Xenia has.

Solving problems with a Cartesian graph

When you understand how to plot points and draw lines, you can use a graph to solve certain types of math problems. When you draw two lines that represent different parts of a word problem, then the point at which the lines *intersect* (where they cross) is your answer.

Here's an example:

> Jacob is exactly 5 years younger than Marnie, and together their ages add up to 15. How old are they?

To solve this problem, first make a chart to show that Jacob is 5 years younger than Marnie:

Jacob	1	2	3	4	5
Marnie	6	7	8	9	10

Then make another chart to show that together, the two children's ages add up to 15:

Jacob	1	2	3	4	5
Marnie	14	13	12	11	10

Finally, plot both lines on a graph (see Figure 12-8) where the horizontal axis represents Jacob's age and the vertical axis represents Marnie's age. Notice that the two lines cross at the point where Jacob is 5 and Marnie is 10 years old, so these are the two children's ages.

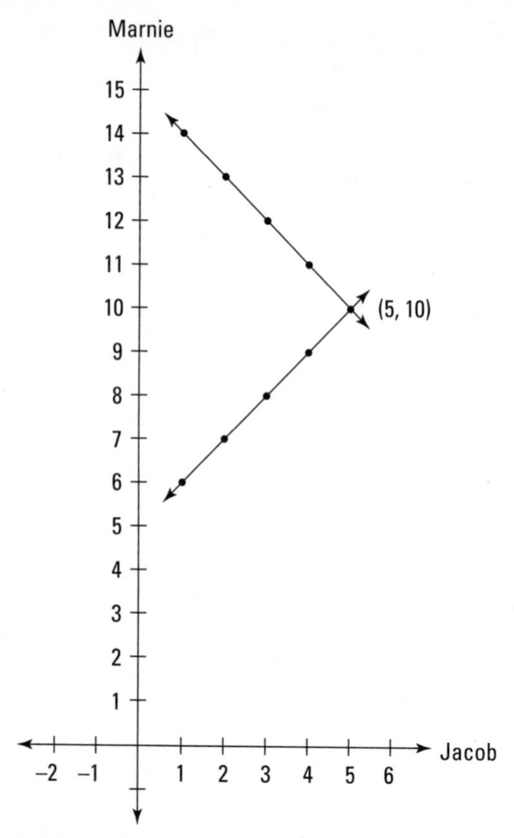

Figure 12-8: Both lines plotted on a graph.

Chapter 13

Ten Essential Math Concepts

In This Chapter

- Appreciating how simple concepts are indispensable in math
- Seeing why π and prime numbers are important
- Understanding the importance of graphs and functions
- Looking at the world of real numbers

Math itself is one big concept, and it's chock full of so many smaller concepts that no one person can possibly understand them all. However, certain concepts get so much airplay that, in my humble opinion, they make the Math Hall of Fame. So here's my list of the ten most important concepts in mathematics.

Playing with Prime Numbers

A *prime number* is any counting number that has exactly two divisors (numbers that divide into it evenly) — 1 and the number itself. Here are the first ten prime numbers:

2 3 5 7 11 13 17 19 23 29 ...

Prime numbers go on forever — that is, the list is infinite. Beyond this, prime numbers are, in an important sense, the elements from which all other numbers can be built. Every counting number greater than 1, no matter how large, can be written as the unique product of prime numbers. See Chapter 1 for more on prime numbers.

Zero: Much Ado about Nothing

Zero may look like a big nothing, but it's actually one of the greatest inventions of all time. Like all inventions, it didn't exist until someone thought of it. The Greeks and Romans, who knew so much about math and logic, knew nothing about zero. The number systems they used had no way to express, for example, how many olive trees you had left when you started with three and an angry neighbor cut down all three of them.

The concept of zero as a number arose independently in several different places. In South America, the number system that the Mayans used included a symbol for zero. And the Hindu-Arabic system, which people use throughout most of the world today, developed from an earlier Arabic system that used zero as a placeholder.

Delicious Pi

The symbol π (pi — pronounced *pie*) is a Greek letter that stands for the ratio of the circumference of a circle to its diameter (see Chapter 11 for the scoop on circles). Here's the approximate value of π:

$$\pi \approx 3.1415926535...$$

Although π is just a number — or, in algebraic terms, a *constant* — it's important for several reasons:

- Geometry just wouldn't be the same without it. Circles are one of the most basic shapes in geometry, and you need π to measure the area and the circumference of a circle. So if you just want to know the area of your round kitchen table, π can come in handy.
- Pi is an *irrational number,* which means that no fraction equals it exactly. Even though π emerges from a very simple operation (measuring a circle), it contains a deep complexity that numbers such as 0, 1, –1, $\frac{1}{2}$, and even $\sqrt{2}$ don't share.

- Pi is everywhere in math. It shows up constantly (no pun intended) where you'd least expect it. One example is trigonometry, the study of triangles. Triangles obviously aren't circles, but trig uses circles to measure the size of angles, and you can't swing a compass without hitting π.

Equal Signs and Equations

Almost everyone takes the humble equal sign (=) for granted. It's so common in math that it goes virtually unnoticed. But the fact that the equal sign shows up practically everywhere only adds weight to the idea that the concept of *equality* — an understanding of when one thing is mathematically the same as another — is one of the most important math concepts ever created.

A mathematical statement with an equal sign is an *equation*. The equal sign links two mathematical expressions that have the same value. The power of math lies in this linkage. That's why nearly everything in math involves equations. On their own, expressions are limited in their usefulness. The equal sign provides a powerful way to connect expressions, which allows scientists to connect ideas in new ways.

The Cartesian Graph

The Cartesian graph (also called the Cartesian coordinate system) is the fancy name for the good old-fashioned x, y plane, which I discuss in Chapter 12. It was invented by French philosopher and mathematician René Descartes.

Descartes's invention of the graph brought algebra and geometry together. The result was *analytic geometry*, a new mathematics that not only merged the ancient sciences of algebra and geometry but also brought greater clarity to both. Now you can draw solutions to equations that include the variables x and y as points, lines, circles, and other geometric shapes on a graph.

Relying on Functions

A *function* is a mathematical machine that takes in one number (called the *input*) and gives back exactly one other number (called the *output*). It's kind of like a blender, because what you get out of it depends on what you put into it.

Suppose I invent a function called PlusOne that adds 1 to any number. So when you input the number 2, the number that gets outputted is 3:

PlusOne(2) = 3

Similarly, when you input the number 100, the number that gets outputted is 101:

PlusOne(100) = 101

As you can see, whenever you input an even number, the function PlusOne outputs an odd number. And this will happen for every even number. Thus, this function maps the set of even numbers onto the set of odd numbers.

Functions get a lot of play as you move forward in algebra. For now, just remember that a function takes an input and gives you an output. For a deeper look at functions, see *Algebra For Dummies* by Mary Jane Sterling (Wiley).

Rational Numbers

The *rational numbers* include the integers and all the fractions between the integers. Here, I list only the rational numbers from –1 to 1 whose denominators (bottom numbers) are positive numbers less than 5:

$$\ldots -1 \ldots -\frac{3}{4} \ldots -\frac{2}{3} \ldots -\frac{1}{2} \ldots -\frac{1}{3} \ldots -\frac{1}{4} \ldots 0 \ldots \frac{1}{4} \ldots \frac{1}{3} \ldots \frac{1}{2} \ldots \frac{2}{3} \ldots \frac{3}{4} \ldots 1 \ldots$$

The ellipses (. . .) tell you that between any pair of rational numbers are an infinite number of other rational numbers.

Rational numbers are commonly used for measurement in which precision is important. For example, a ruler wouldn't be much good if it were to measure length only to the nearest inch. Most rulers measure length to the nearest $\frac{1}{16}$ of an inch, which is close enough for most purposes. Similarly, measuring cups, scales, precision clocks, and thermometers that allow you to make measurements to a fraction of a unit also use rational numbers.

The set of rational numbers is *closed* under the Big Four operations. That is, if you take any two rational numbers and add, subtract, multiply, or divide them, the result is always another rational number.

Irrational Numbers

In a sense, the *irrational numbers* are a sort of catchall: Every number on the number line that isn't rational is irrational. By definition, no irrational number can be represented as a fraction; nor can an irrational number be represented as either a terminating decimal or a repeating decimal (see Chapter 5 for more about these types of decimals). Instead, an irrational number can be approximated only as a nonterminating, nonrepeating decimal: The string of numbers after the decimal point goes on forever without creating a pattern.

The most famous example of an irrational number is π, which represents the circumference of a circle with a diameter of 1 unit. Another common irrational number is $\sqrt{2}$, which represents the diagonal distance across a square with a side of 1 unit. In fact, all square roots of non-square numbers (such as $\sqrt{3}$, $\sqrt{5}$, and so forth) are irrational numbers.

The Real Number Line

The number line has been around for a very long time, and it's one of the first visual aids that teachers use to teach kids about numbers. Every point on the number line stands for a number. Well, okay, that sounds pretty obvious, but strange to say, this concept wasn't fully understood for thousands of years.

The Greek philosopher Zeno of Elea posed this problem, called *Zeno's Paradox:* In order to walk across the room, you have to first walk half the distance ($\frac{1}{2}$) across the room. Then you have to go half the remaining distance ($\frac{1}{4}$). After that, you have to go half the distance that still remains ($\frac{1}{8}$). This pattern continues forever:

$$\frac{1}{2} \quad \frac{1}{4} \quad \frac{1}{8} \quad \frac{1}{16} \quad \frac{1}{32} \quad \frac{1}{64} \quad \frac{1}{128} \quad \frac{1}{256} \cdots$$

So you can never get to the other side of the room.

Exploring the Infinite

The very word *infinity* commands great power. So does the symbol for infinity (∞). How big is infinity? Here's a common answer: If you were to count all the grains of sand on all the beaches in the world and then do the same thing on every planet in our galaxy, by the time you were done counting, you'd be no closer to infinity than you are right now. In fact, infinity isn't a number at all. *Infinity,* beyond any classification of size or number, is the very quality of endlessness. And yet, mathematicians have tamed infinity to a great extent.

Index

• Symbols and Numerics •

• A •

• B •

• C •

• D •

• E •

• F •

• G •

• H •

• I •

• L •

• M •

• Q •

• R •

• S

• T •

• V •

• W •

• X •

• Z •